Cryopedology

Progress in Soil Science

Aims and Scope

Progress in Soil Science series aims to publish books that contain novel approaches in soil science in its broadest sense – books should focus on true progress in a particular area of the soil science discipline. The scope of the series is to publish books that enhance the understanding of the functioning and diversity of soils in all parts of the globe. The series includes multidisciplinary approaches to soil studies and welcomes contributions of all soil science subdisciplines such as: soil genesis, geography and classifi cation, soil chemistry, soil physics, soil biology, soil mineralogy, soil fertility and plant nutrition, soil and water conservation, pedometrics, digital soil mapping, proximal soil sensing, soils and land use change, global soil change, natural resources and the environment.

More information about this series at http://www.springer.com/series/8746

James G. Bockheim

Cryopedology

James G. Bockheim
Department of Soil Science
University of Wisconsin
Madison, WI, USA

ISBN 978-3-319-08484-8 ISBN 978-3-319-08485-5 (eBook)
DOI 10.1007/978-3-319-08485-5
Springer Cham Heidelberg New York Dordrecht London

Library of Congress Control Number: 2014952316

Printed on acid-free paper

Springer is part of Springer Science+Business Media (www.springer.com)

Preface

I became interested in cryopedology while a beginning graduate student at the University of Maine studying the properties and genesis of alpine soils (1967–1968) in the Appalachian Mountains. This interest continued as I studied the genesis of alpine and subalpine soils in the North Cascades Mountains (1968–1972) under the guidance of Fiorenzo Ugolini at the University of Washington. While at Washington, I was invited by Fio for my first trip to Antarctica in 1969. I spent parts of the summer of 1973 with Tim Ballard (University of British Columbia) on Cornwallis Island and 1974 with Art Dyke (then a PhD student at the University of Colorado) on Baffin Island, Nunavut, in the Canadian Arctic Archipelago. From 1975 to 1987, I used soils to date and correlate glacial deposits in the Transantarctic Mountains with George Denton of the University of Maine. During 1995–2007, I had the good fortune to study soil development and distribution in arctic Alaska with Donald ("Skip") Walker, Ken Hinkel, and Frederick ("Fritz") Nelson. I spent parts of three summers (1992, 1997, 1998) studying soil development and glacial geomorphology in the High Uintas of Utah as a guest of former student Darlene Koerner of the Ashley National Forest. I was invited by Malcolm McLeod and Jackie Aislabie of Landcare Research, New Zealand, to return to Antarctica; we worked together from 2004 through 2013 mapping soils as part of an environmental domains project. In 2011, Gonçalo Vieira invited me to participate in his PERMANTAR research project in the South Shetland Islands, maritime Antarctica. The US National Science Foundation supported me from 2011 to 2014 to study permafrost, active-layer dynamics, and soils of the western Antarctic Peninsula. In all, I've spent 19 field seasons in Antarctica, 15 in the arctic, and 9 in alpine areas.

My interest in writing a book on cryopedology was spurred by the excellent but unfortunately now outdated works of J.C.F. Tedrow, *Soils of the Polar Landscapes* (1977), and Iain Campbell and Graeme Claridge, *Antarctica: Soils, Weathering and Environment* (1987). *Cryosols: Permafrost-affected Soils* (2004), edited by John M. Kimble, contributed substantially to our understanding of arctic and Antarctic soils. In 2014, I was asked to edit *Soils of Antarctica* as part of Springer's World Soils Book Series. That same spring, Carlos E.G.R. Schaefer invited me to teach a short course on cryopedology at the Federal University of Viçosa in Brazil. I found that this course required me to collate information in the previously mentioned works with that in A.L. Washburn's *Periglacial Processes and Environments* (1973) and Hugh French's *The Periglacial Environment* (2007) as well as a number of published papers that are cited in the present volume.

I hope that this book is of use to pedologists, geographers, geologists, and others who are fascinated by soils of the cold regions.

Madison, WI, USA James G. Bockheim

Acknowledgments

The idea for this book was inspired by Carlos Schaefer and Felipe Simas, who graciously invited me to their university, the Federal University of Viçosa in Brazil, to collaborate with them on their research and to teach a short course on cryopedology. I am also indebted to my fellow "cryopedologists," in particular my late colleague David Gilichinsky, who hosted the first International Meeting on Cryopedology in 1992 in Pushchino, Moscow Region, Russia. I have enjoyed my collaborations with Megan Balks, Hans-Peter Blume, Manfred Bölter, Gabrielle Broll, Jerry Brown, Iain Campbell, Graeme Claridge, Roman Desyatkin, Kaye Everett, Sergey Goryachkin, Cezary Kabala, John Kimble, Dmitri Konyushkov, Malcolm McLeod, Yevgeny Naumov, Eva-Maria Pfeiffer, Chien-Lu Ping, Ilya Sokolov, Charles Tarnocai, John Tedrow, Fio Ugolini, and other colleagues.

As always, my wife Julie offered me encouragement in this endeavor, and it is to her the book is dedicated.

Contents

Chapter 1
Introduction

1.1 History

The term "cryopedology" is derived from the Greek words *cryos* ("icy cold"), *pedon* ("soil"), and *logos* ("study") and, hence, refers to the study of frozen ground and intensive frost action. Although cryosols were studied for many years in Russia, Nikiforoff (1928) introduced the concept of frozen ground and intensive frost action to the English-speaking world. He provided a historical overview of permafrost, a map showing the distribution of permafrost in Eurasia, a summary of data regarding permafrost thickness, and its relation to present-day and paleo-climates. Kirk Bryan (1946), the American geomorphologist, introduced "cryopedology" as the study of frozen ground and intensive frost action. In 1949, the French geomorphologist André Guilcher traced the development of cryopedology. Cailleux and Taylor (1954) published *Cryopedology: the Study of Frozen Soils* as part of the French polar expedition to Greenland.

Makeev (1999) discussed the history and aspects of cryopedology and provided an overview of the genesis, classification, geography, and management of cryogenic soils. Some historical highlights of cryopedology are provided in Table 1.1. The earliest formal observations of permafrost (1864) seem to be those of the Russian naturalist A. Middendorf while traveling in Russia. Considered by many to be the "father" of pedology, Dokuchaev (1883) identified the arctic soil zone in Russia. The Germans E. Ramann, W. Meinardus, and E. Blanck studied soils on Spitzbergen and Greenland in 1911–1912. While on the Shackleton Expedition of 1907–1909, Jensen (1916) collected and analyzed soil samples from Ross Island. Another famous pedologist, K.K. Glinka, led a soils expedition to Siberia in 1921. M.I. Sumgin (1927) emphasized the effects of permafrost on soil processes; in 1934, Y.A. Livorskiy published a book on tundra soils of northern Russia (cited by Goryachkin et al. 2004).

The first investigations of permafrost-affected soils in Canada were led by A. Leahey in the middle to late 1940s (Table 1.1). Supported by the U.S. Naval Research Laboratory, John C.F. Tedrow began his studies of permafrost-affected

J.G. Bockheim, *Cryopedology*, Progress in Soil Science,
DOI 10.1007/978-3-319-08485-5_1

Table 1.1 Some highlights in the history of cryopedology

Year	Scientist, country	Event
1864	Middendorf, Germany	Investigated permafrost in Siberia
1900	Dokuchaev, Russia	Identified arctic soil zone in Russia
1911	Ramann, Germany	Compared soils of Spitzbergen and central Asian plateau
1912	Meinardus, Blanck, Germany	Studied soils on Spitzbergen
1916	Jensen	First study of soils in Antarctica
1921	Glinka, Russia	Soils expedition to Siberia
1927	Sumgin, Russia	Permafrost effects on soil processes
1934	Livorskiy, Russia	Book: *Tundra soils of the Northern Region*
1943	Leahey, Canada	First studies of soils in northern Canada
1951	Kellogg, Nygard, USA	Exploratory study of soils in arctic Alaska
1956	Markov, Russia	Studied soils at Mirnyy, East Antarctica
1960	McCraw, New Zealand	First soil map in Antarctica, Taylor Valley
1962	Ivanova, Russia	Books: *On Soils of Siberia*
1966	Tedrow, USA	Book: *Antarctic Soils and Soil Forming Processes*
1977	Tedrow, USA	Book: *Soils of the Polar Landscapes*
1978	Tarnocai, Canada	Establishment of Cryosol order Canadian soil taxonomy
1983	Rieger 1983	Book: *The Genesis and Classification of Cold Soils*
1987	Campbell, Claridge, NZ	Book: *Antarctica: Soils, Weathering Processes & Environment*
1992	Gilichinsky, Russia	First International Conference on Cryopedology
1993	Tarnocai, Canada; Ping, USA	Field trip permafrost-affected soils of AK, Yukon, TWT
1999	Bockheim, USA	Gelisol order accepted in *Soil Taxonomy*
2004	Kimble, USA	Book: *Cryosols: Permafrost-Affected Soils*
2011	Jones et al.	Book: *Soil Atlas of Northern Circumpolar Region*
2014	Bockheim, USA	Book: *Soils of Antarctica*

soils in the North American arctic in the early 1950s. Tedrow's work culminated in the publication of *Soils of the Polar Landscapes* in 1977. In 1978 the cryosol order was introduced into the Canadian soil classification system, spurred by the efforts of Charles Tarnocai. Samuel Rieger mapped soils in the permafrost region of Alaska with the Natural Resource Conservation Service (then the Soil Conservation Service) and published *The Genesis and Classification of Cold Soils* in 1983. Although this book focused on Alaska, it addressed temperature relations in cold soils, the effects of freezing, and the use of *Soil Taxonomy* in classifying soils of the cold regions.

In 1987 Iain Campbell and Graeme Claridge published *Antarctica: Soils, Weathering Processes and Environment*, which has been the most comprehensive treatment of Antarctic soils for over 25 years (Table 1.1). The Gelisol order was accepted into *Soil Taxonomy* in 1999 under the leadership of James Bockheim. In 2004 John Kimble edited *Cryosols*: *Permafrost-affected Soils*, which contained

Table 1.2 A listing of international conferences on cryopedology

No.	Year	Place	Theme	Coordinators
1	1993	Pushchino, Russia	"Cryosols: the effects of cryogenesis on the processes & peculiarities of soil formation"	David Gilichinsky
2	1997	Syktyvkar, Russia	"Cryogenic soils: ecology, genesis, classification"	Galina Mazhitova, David Gilichinsky
3	2001	Copenhagen, Denmark	"Dynamics & challenges of cryosols"	Bjarne Jakobsen
4	2005	Arkangelsk, Pinega, Russia	"Cryosols: genesis, ecology & management"	Sergey Goryachkin
5	2009	Ulan-Ude, Russia	"Diversity of frost-affected soils & their role in ecosystems"	Sergey Goryachkin
6	2013	Krakow, Poland	"Frost-affected soils: dynamic soils in a dynamic world"	Marek Drewnik

37 papers describing the history, geography, properties, processes, classification, and management of cryosols in the arctic and Antarctic regions. In 2014 Bockheim edited *Soils of Antarctica* that included chapters discussing soils of each of 10 ice-free regions in Antarctica. More detailed analyses of the history of cold soils research are given in Tedrow (1977, 2004), Goryachkin et al. (2004), and Tarnocai (2004).

An interest in the expanding sub-discipline of cryopedology led to a series of six International Conferences on Cryopedology which have been held approximately every 4 years, beginning in 1993; the last meeting was held in 2013 (Table 1.2). The meetings have drawn 60 or more participants each from the international cryopedological community. However, full papers were published only from the original 1993 meeting.

The term "cryopedology" has not attained much recognition in the international literature. A search of the Web of Science database yielded only three citations since the 1990s. However, "soils of polar regions" yielded nearly 400 citations, with an exponential increase beginning in 1995 (Fig. 1.1). The sub-science of cryopedology has gained considerable recent attention because of the effects of high-latitude and high-elevation warming on release of CO_2. However, to date a book introducing the basics of cryopedology has not been published.

1.2 Soil Concept

Up until 1999 the standard definition of the soil in the USA was "a collection of natural bodies … containing living matter and supporting or capable of supporting plants out-of-doors" (Soil Survey Staff 1975, p. 1). Questioned as to whether the weathered unconsolidated materials in Antarctica were "soils," Bockheim (1982, p. 240) provided a working definition of soil as "a natural body comprised of solids

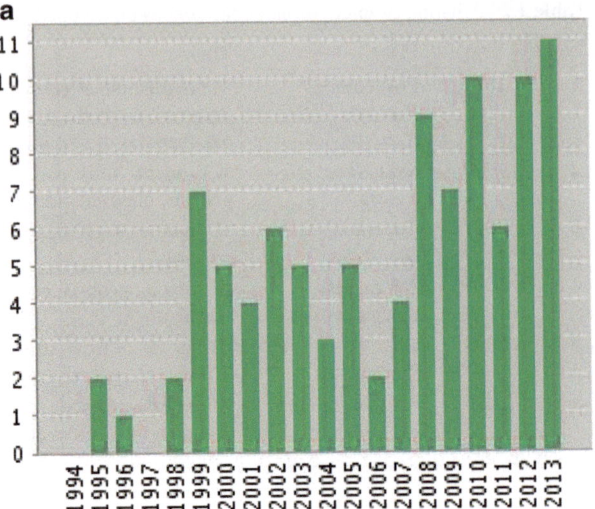

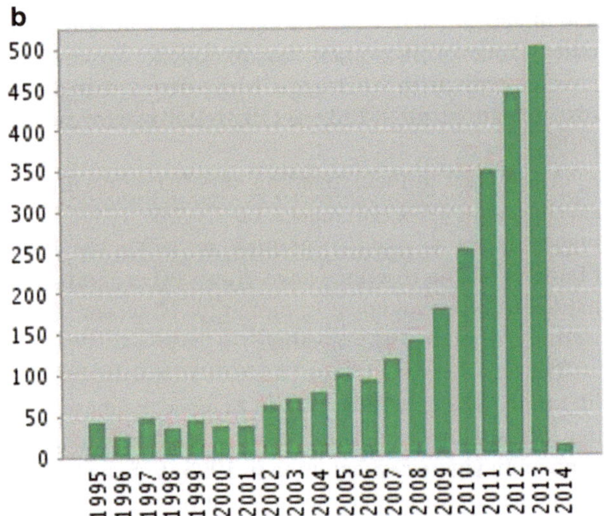

Fig. 1.1 (**a**) Publications in and (**b**) total citations to "soils of polar regions" in the Web of Science 1995 through 2013

… liquid … and gases … which occurs on the land surface, occupies space, and is organized into horizons which are readily distinguishable from the initial material as a result of additions, losses, transfers, and transformation of energy and matter." Inasmuch as cryptogams and microbiota are present on and within the soils of Antarctica, he argued that Antarctic soils "supported plants out-of-doors." By the second edition of *Soil Taxonomy* (Soil Survey Staff 1999), the soils community had accepted his definition in recognition of soils in Antarctica.

The term "cryosol" is widely accepted in the cryopedological literature and has been adopted as a key soil group in the *World Reference Base for Soil Resources*

(IUSS Working Group WRB 2006) and soil order in the *Canadian System of Soil Classification* (Soil Classification Working Group 1998). The term is consistent with the use of terms such as "cryoturbation" (soil movements due to frost action), "cryosphere" (that portion of the Earth where ice and snow occurs), "cryostructures" (structures present in frozen materials), etc. When permafrost-affected soils were being incorporated into *Soil Taxonomy*, the term "Gelisol" was used in order to construct pronounceable soil taxonomic names. The tradition in ST was to build these names from the first vowel and the consonant that followed. This was not possible with the term "cryosol" but "el" could be used to construct soil taxonomic categories. This term is consistent with the use of terms such as "gelifluction" (soil creep associated with permafrost) and "gelic materials" (soil materials influenced by cryogenic processes).

Finally, there has been a proliferation of journals specializing in cryopedology and related studies, including *Arctic Journal* (began in 1948), *Arctic, Antarctic and Alpine Research* (1969), *Antarctic Science* (1989), *Cold Regions Research and Technology* (1979), *Permafrost and Periglacial Processes* (1990), *Polar Biology* (1982), *Polar Record* (1931), and *Polar Science* (2007), as well as the mainline soils journals.

References

Bockheim JG (1982) Properties of a chronosequence of ultraxerous soils in the Trans-Antarctic Mountains. Geoderma 28:239–255

Bockheim JG (ed) (2014) Soils of Antarctica, World soils book series. Springer, New York

Bryan K (1946) Cryopedology: the study of frozen ground and intensive frost action, with suggestions on nomenclature. Am J Sci 244:622–642

Cailleux A, Taylor G (1954) Cryopedology: the study of frozen soils. Hermann, Paris

Campbell IB, Claridge GGC (1987) Antarctica: soils, weathering and environment. Elsevier, New York

Dokuchaev, VV 1883 The Russian Chernozem Report to the Free Economic Society (in Russian). Imperial Univ. of St. Petersburg, St. Petersburg, Russia

Goryachkin SV, Karavaeva NA, Makeev OV (2004) The history of research of Eurasian cryosols. In: Kimble JM (ed) Cryosols: permafrost-affected soils. Springer, New York, pp 17–28

Guilcher A (1949) Le developpement de la cryopédologie. Annales de Géographie 58:336–338

IUSS Working Group WRB (2006) World reference base for soil resources 2006. World soil resources reports no. 103. FAO, Rome

Jensen HI (1916) Report on Antarctic soils. Reports on science investigation on British Antarctic expedition 1907–1909 Part IV. Geology 2:89–92

Kimble JM (ed) (2004) Cryosols: permafrost-affected soils. Springer, Berlin

Makeev OV (1999) Soil, permafrost, and cryopedology. Eurasian Soil Sci 32:854–863

Nikiforoff CC (1928) The perpetually frozen subsoil of Siberia. Soil Sci 26:61–81

Rieger S (1983) The genesis and classification of cold soils. Academic, New York

Soil Classification Working Group (1998) The Canadian system of soil classification, 3rd edn, Research Branch, Agricultural & Agri-Food Canada. NRC Research Press, Ottawa

Soil Survey Staff (1975) Soil taxonomy: a basic system of soil classification for making and interpreting soil surveys, Agricultural handbook no. 436. USDA Soil Conservation Service, Superintendent of Documents, U.S. Government Printing Office, Washington, DC

Soil Survey Staff (1999) Soil taxonomy: a basic system of soil classification for making and interpreting soil surveys, 2nd edn, Agricultural handbook no. 436. USDA Soil Conservation Service, Superintendent of Documents, U.S. Government Printing Office, Washington, DC

Sumgin MI 1927 Permafrost in USSR. Far-East Geophysical Observatory: Vladivostok; 369 pp (In Russian).

Tarnocai C (2004) Northern soil research in Canada. In: Kimble JM (ed) Cryosols: permafrost-affected soils. Springer, New York, pp 29–43.

Tedrow JCF (1977) Soils of the polar landscapes. Rutgers University Press, New Brunswick, New Jersey

Tedrow JCF (2004) Soil research in arctic Alaska, Greenland, and Antarctica. In: Kimble JM (ed) Cryosols: permafrost-affected soils. Springer, New York, pp 5–16

Chapter 2
Cryosols as a Three-Part System

2.1 Three-Part Conceptual Model of Cryosols

The standard conception of permafrost-affected soils employs a simple two-layer model. At depth is permafrost, which is defined as "earth material that remains continuously at or below 0 °C for at least two consecutive years" (van Everdingen 1998). At the surface is the "active layer," which thaws in summer and refreezes in winter. Recent research in North America supports an earlier Russian proposal for the existence of a "transition" or "transient" layer in polar soils (Shur et al. 2005). This layer—the uppermost, or near-surface, portion of the permafrost—meets the thermal criteria for permafrost defined above (Fig. 2.1). However, its maximum summer thaw depth tends to fluctuate from year to year owing to interannual variations in components of the surface energy budget. During occasional deep thaw penetration, ice in the near-surface permafrost melts, and the affected zone temporarily becomes part of the active layer. Therefore, it is not unusual for the soil profile to extend into the near-surface permafrost (Fig. 2.2).

2.2 Active Layer

2.2.1 *Definition*

The active layer is the layer of ground that is subject to annual thawing and freezing in areas underlain by permafrost. In the zone of continuous permafrost, the active layer generally reaches the permafrost table, but in the zone of discontinuous permafrost, it may not. Some scientists consider the active layer to include the uppermost part of the permafrost. However, this is referred to this as the transition zone or transitory zone in this book (Fig. 2.1).

J.G. Bockheim, *Cryopedology*, Progress in Soil Science,
DOI 10.1007/978-3-319-08485-5_2

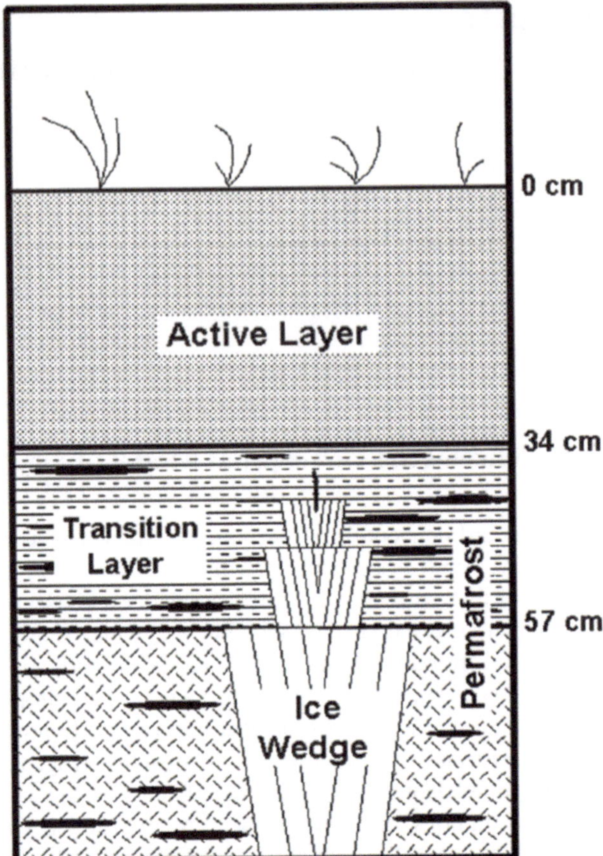

Fig. 2.1 Schematic diagram of a three-layer conceptual model. Note formation of ice lenses (*black ellipses*), upward growth of secondary and tertiary ice wedges into the transition zone, and an ice vein protruding from tertiary ice wedge (Bockheim and Hinkel 2005)

Fig. 2.2 A cryoturbated soils from arctic Alaska. The flag pins denote the end-of-season permafrost table. An organic horizon and streaks of organic-rich mineral soil occur below the permafrost table (Photo by J. Bockheim)

2.2.2 *Dynamics and Modeling the Active Layer*

There have been many attempts to model the active-layer thickness from climate-permafrost interactions (e.g., Mackay 1995). As an alternative to these approaches, semi-empirical methods have been developed for the practical needs of cold-regions engineering. These include the "frost index" (Nelson and Outcalt 1987), the "n-factor" (Klene et al. 2001), and Kudryavtsev's formulation (Anisimov et al. 1997). The "frost index" is a dimensionless ratio defined by manipulation of either freezing-and-thawing degree-day sums or frost and thaw penetration depths for detecting the presence of permafrost. The n-factor is defined as the ratio of the sum of thawing degree days at the soil-surface to that in the air. Kudryavtsev's solution takes into account the effects of snow-cover, vegetation, soil moisture, and soil thermal properties and has been widely used to estimate active-layer thickness. The Basal Temperature of Snow (BTS) has been used to predict the occurrence of isolated permafrost in mountainous areas (e.g., Lewkowicz and Ednie 2004).

2.2.3 *Properties of the Active Layer*

2.2.3.1 Thickness

Active layer thickness depends on terrain factors such as air temperature, vegetation, drainage, soil or rock type, texture, total water content, snow-cover, and degree and orientation of slope. The active layer tends to be thin (0.1–0.15 m) in high latitude areas of the zone of continuous permafrost, such as the High Arctic and at high elevations in the Transantarctic Mountains (Table 2.1). In alpine regions of the world, the active layer commonly ranges from 2 to 8 m or more. In the definition of Gelisols, the active layer can not exceed 1 m for Histels (organic cryosols) and Orthels (mineral cryosols without cryoturbation) or 2 m for Turbels (cryoturbated mineral soils). Although these depths are hypothetical, they were chosen as the maximum expression of permafrost effects in terms of soil development, response to climate change, and engineering properties. However, in many mountainous areas where the permafrost is sporadic or isolated, the active layer may extend to depths of 8 m or more (Bockheim and Munroe 2014).

2.2.3.2 Physical Properties

Because the active layer involves the merging of freezing fronts during "freeze-back" (seasonal refreezing of the thawed active layer), it is subject to compression and rearrangement of coarse fragments and the soil matrix that results in changes in physical properties such as structure, bulk density, dilatancy, and texture. Soil horizons in the active layer have granular, platy, prismatic, and blocky structures because of freeze-thaw processes (Tarnocai and Bockheim 2011). Massive structures associated with high bulk density result from repeated cryostatic desiccation, which

Table 2.1 Active-layer depths and permafrost temperatures for selected stations in Antarctica, the Arctic, and alpine regions with permafrost

Station	Latitude (°)	Longitude (°)	Elev. (m)	Active-layer depth (m)	Permafrost temp. (°C)
Antarctica					
Troll	72.011S	2.533E	1,335	0.08	−17.8
Sanae	71.687S	2.842W	805	0.15	−16.8
Novozalarevskaya	70.763S	11.795E	80	0.7	−9.7
Farjuven Bluffs	72.012S	3.388W	1,220	0.25	−17.8
Syowa	69S	39.583E	15	nd	−8.2 (6.8)
Molodezhnaya	66.275S	100.760E	7	0.9–1.2	−9.8
Progress	69.404S	76.343	96	>0.5	−12.1
Larsemann Hills	69.4S	76.27E	50	1.0–1.1	–
Casey Station	66.28S	110.52E	10–100	0.3–0.8	–
Oasis	74.7S	164.100E	80	1.6	−13.5
Boulder clay	74.746S	164.021E	205	0.25	−16.9
Marble Point	77.407S	163.681E	85	0.4	−17.4
Victoria Valley	77.331S	161.601E	399	0.24	−22.5
Bull Pass	77.517S	161.850E	150	0.5	−17.3
Minna Bluff	78.512S	166.766E	35	0.23	−17.4
Scott Base	77.849S	166.759E	80	0.30	−17.0
Ellsworth Mtns.	78.5S	85.6W	800–1,300	0.15–0.50	–
Russkaya	74.763S	136.796	76	0.1	−10.4
Signy Island	60.7S	45.583W	90	0.4–2.2	−2.4
King George Island	62.088S	58.405W	37	1.0–2.0	−0.3 to −1.2
Deception-Livingston Is.	62.67S	60.382W	272	1.0	−1.4 to −1.8
Cierva Point	64.15S	69.95W	182	2.0–6.0	−0.9

Amsler Island, Palmer	64.77S	64.067W	67	14.0	–0.2
Rothera	67.57S	68.13W	32	1.2	–3.1
Marambio Station	64.24S	56.67W	5–200	0.6	–
Arctic					
Northern Norway	69N	18–25E	585–990	>7–>10	0.2 to –0.3
Svalbard	78–79N	13–16E	9–464	0.8–2.5	–2.3 to –5.3
Northern Sweden	68N	18–21E	355–1,550	0.5–1.6	–0.1 to –2.4
Northern Finland	70N	27E	290	0.6	0 to –0.2
Iceland	64–65N	14–18E	899–931	2.0–10	0 to –0.5
West Greenland	64–69N	51–53E	26–49	0.9	0.2
East Greenland	74N	20W	37	0.8–1.5	–1.1 to –8.1
Alaska lowland	60–66N, 69–73N	143–180W	5–500	0.35–0.60	0 to –5.0
Western Canada	58–69N	120–140W	5–500	0.6–1.9	0 to –8.0
Western No. Amer. Cordillera	60–65N	125–140W	>2,180–>3,000	1.5–3.5	0 to –0.5
Central Canada	57–75N	92–95W	5–500	3.5–7.0	0 to –12
Eastern Canada	55–82.5N	92–63W	5–500	1.4–5.6	0 to –10
Bolvansky Cape, Russia	68N	54E	25–30	0.3–5.0	0.5 to –1.9
Vorkuta, Russia	63–64N	67–68E	20–154	0.3–1.4	–0.2 to –2.4
Nadym, Russia	73N	65E	25	0.3–1.4	–0.04 to –0.52
Urengoy, Russia	77N	66–67E	7	0.3–1.4	–0.04 to –3.9
Northern Yakutia, Russia	69–72N	129–161E	1–43	0.45–1.0	–0.9 to –10.8
Trans-Baikal, Russia	57N	118E	770–1,712		–4.7 to –5.1

(continued)

Table 2.1 (continued)

Station	Latitude (°)	Longitude (°)	Elev. (m)	Active-layer depth (m)	Permafrost temp. (°C)
Alpine					
Western Cordillera, USA	37–49.5, 60–70N	105–115W	>3,500	2.0–>5.0	−1.5 to −2.5
Western Cordillera, Canada	51–60N	115–140W	>2,180–>3,000	–	−0.1 to −1.0
Brooks Range	68–69N	143–161W	>500	0.3–1.5	
Appalachian Mountains	42–45N	71W	>1,200–1,800	>2	–
Andes Mountains, South America	20–57S	70–78W	>1,500–>5,000	3.0–9.0	0
Fennoscandia mountains	63–70N	14–30E	>1,400–>1,800	1.4–3.0	−3.0 to −4.0
Iceland	64–66N	14–23E	>800–>1,000	0.5–0.6	–
Svalbard	77–78N	11–25E	>30	0.9–1.0	−6.3
European Alps	44–47N	6–15E	>2,400–>3,000	0.5–8.0	−0.5 to −2.5
Pyrenees	42N	1W–2E	>2,700	–	–
Carpathians	44–48N	24–27E	>2,100	–	–
Urals	50–60N	58–60E	>550–>	0.3–0.5	–
Caucasus	41–43N	40–48E	>2,800–>3,000	–	–
Himalayas-Karakoram-Hindu Kush (Qinghai-Tibet Plateau)	30–40N	70–105E	>4,600–>5,000	1.0–7	−0.5 to −3.5
Altai Mtns., Mongolia	45–53N	87–97E	>2,400–>3,600	1.0–10.0	−0.14 to −1.6
Pamir-Tien Shan-Djunder Alatau	40–44N	69–95E	>2,700–>3,000	0.5–4	−0.3 to −6.7
Yablonoi-Sayan-Stanovoi Mtns., Siberia	50–52N	92–120E	>1,750	>0.5	
Japanese Alps	44N	143E	>1,600–>2,000	1.0–5.5	–

The table will appear in Bockheim and Munroe (2014) with all of the references develops as a result of the movement of two freezing fronts (downward from the surface and upward from the permafrost table) during freeze-back periods. These two freezing fronts draw water from the unfrozen soil layer between the two fronts and compact this layer because of the cryostatic pressure they exert.

Fine-textured cryosols commonly have high moisture content, especially above the permafrost layer; resulting in gleying and other redoximorphic features. Salt crusts also form on the surface of cryosols, especially those developed from marine clays and shale-derived parent materials.

Dilatancy is common in cryosols with high silt content. This property leads to the creation of a very unstable soil surface that liquefies when subjected to mechanical vibration. When these soils dry, a characteristic vesicular structure develops.

2.2.3.3 Chemical Properties

Cryoturbation within the active layer has marked effects on chemical properties such as the distribution and turnover of soil organic C and N, Fe and Mn redox relations, hydrogeochemistry of seasonal flow regimes (Pecher 1994), and epigenic salt accumulation. The effect of sporadic permafrost on chemical soil properties was examined in the eastern Swiss Alps (Zollinger et al. 2013). Although there were no significant differences in C stocks on permafrost and non-permafrost sites, the stable C-fraction was greater in the non-permafrost sites, enabling greater turnover of SOC.

One of the unique properties of cryosols, especially those affected cryoturbation, is the large amount of SOC in both the active and transitional layers (Bockheim et al. 1998). Even though permafrost-affected ecosystems produce much less biomass than do temperate ecosystems, permafrost-affected mineral soils that are subject to cryoturbation are able to sequester a large portion of this organic matter (Bockheim 2007). Although permafrost-affected mineral soils cover a much greater area than permafrost-affected organic soils, the organic cryosols are able to sequester SOC at much greater rates as a result of their gradual build-up process.

2.2.3.4 Mineralogical Properties

Putkonen (1998) pointed out that the thermal regime of the active layer is important because all chemical, biological and physical processes are concentrated there. Moreover, thermal conduction, the phase change of soil water at 0 °C, and changes in unfrozen water content are the primary thermal processes that control soil temperature and weathering rates. Alekseev et al. (2003) compared mineral transformations in the active layer versus that in the near-surface permafrost. Illite and dioctahedral (Fe) chlorite were altered at the cryogenic barrier to lepidocrocite. They identified the boundary between the active layer and permafrost as an important geochemical barrier. Borden et al. (2010) compared the clay mineralogy of soils in moist acidic tundra (MAT) and moist nonacidic tundra (MNT) of arctic Alaska. They reported that the proportion of vermiculite to illite was higher in MAT than in MNT because of differences in acidity and its relation to weathering processes.

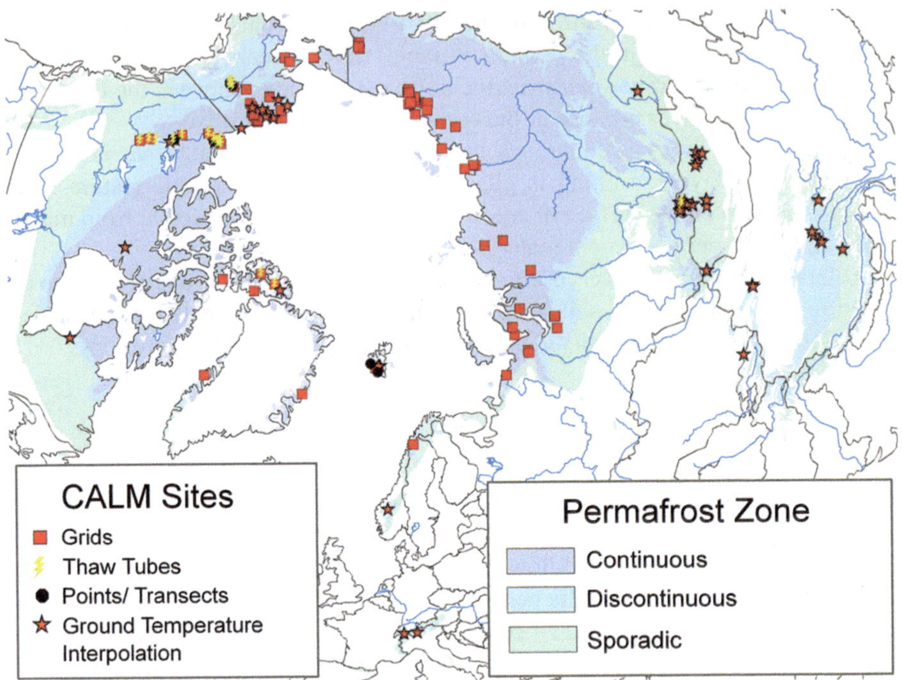

Fig. 2.3 CALM sites in the circumarctic (Source: CALM; http://www.gwu.edu/~calm)

2.2.4 Circumpolar Active Layer Monitoring

Because of the importance of the active layer in recording climate change, active-layer monitoring networks have been established in the circumarctic and Antarctic regions (http://www.gwu.edi/~calm). The Circumpolar Active Layer Monitoring (CALM) network was established in the 1990s and observes the long-term response of the active layer and near-surface permafrost to changes and variations in climate. The CALM network includes more than 200 sites and has participants from more than 15 countries. CALM sites in the northern and southern hemispheres are shown in Figs. 2.3 and 2.4, respectively.

2.3 Transient Layer

The standard conception of permafrost-affected soils employs a simple two-layer model. At depth is permafrost, which is defined as "*earth material that remains continuously at or below 0 °C for at least two consecutive years*" (van Everdingen 1998). At the surface is the "active layer" which thaws in summer and refreezes in winter. However, recent research in North America supports an earlier Russian proposal for the existence of a "transition" or "transient" layer

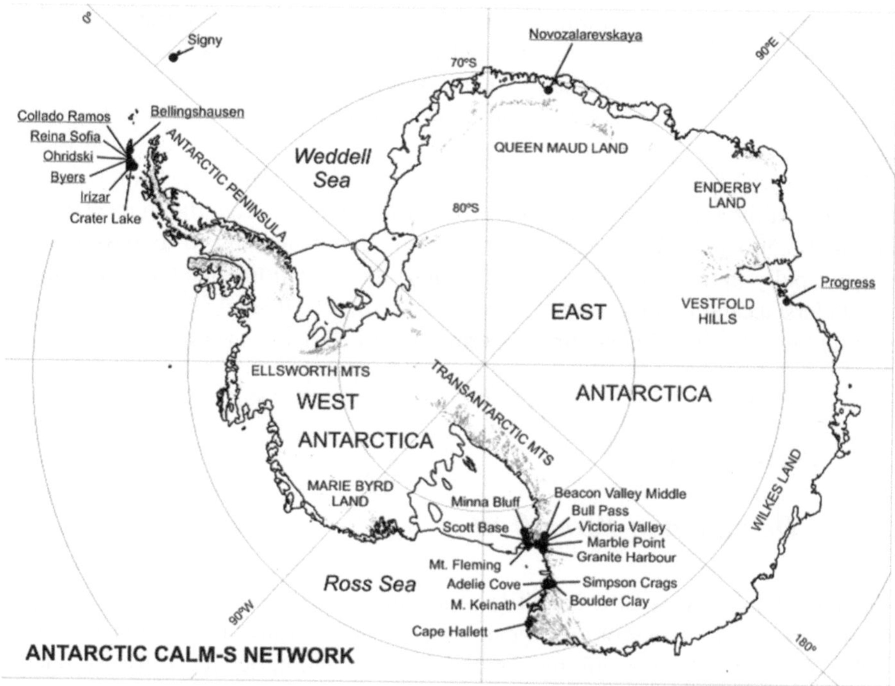

Fig. 2.4 Antarctic circumpolar monitoring network (CALM-S) (Vieira et al. 2010)

in arctic soils (Shur et al. 2005; Bockheim and Hinkel 2005). This layer is the uppermost, or near-surface, portion of the permafrost and meets the thermal criteria defined above.

However, the maximum summer thaw depth tends to fluctuate from year to year owing to interannual variations in components of the surface energy budget. During occasional deep thaw penetration, ice in the near-surface permafrost melts and the affected zone temporarily becomes part of the active layer. It is expected that these episodic thaw events will occur at decreasing frequencies with greater depth; stated another way, it suggests that the thaw recurrence interval increases with depth within the transition zone (Shur et al. 2005). On a time scale ranging from sub-decadal to multi-centennial, all or some of this zone temporarily thaws. This is viewed as a response to both interannual variation and possibly to longer-term climate changes. The base of the transition zone, therefore, marks the position of maximum thaw over the time interval; this can be considered the long-term permafrost table.

The affected portion of the transition zone returns to the frozen state following episodic deep thaw. Owing to its unique history, it has characteristics that differ from both the active layer above and the deeper permafrost below. As a result, the transition zone has a large impact on soil and cryogenic structures and on the thermal stability of permafrost.

The transition zone exhibits the effects of cryoturbation, contains abundant redistributed organic carbon, is enriched in ice in the forms of lenses, veins and reticulate vein ice (nets), and has abundant soil moisture. In Arctic Alaska, the surface of the transition zone was found at an average depth of 34 ± 7 cm below the ground surface and had an average thickness of 23 ± 8 cm (Bockheim and Hinkel 2005). They observed no significant differences in the depth of the boundaries and thickness of the transition zone in drained thaw-lake basins ranging in age between 300 and 5,500 year BP, suggesting that the processes leading to its development occur rapidly in arctic Alaska. Recognition of the transition zone has implications for understanding pedogenic processes in permafrost-affected soils and for determining the response of near-surface permafrost to climate warming.

2.4 Permafrost

2.4.1 Definition of Permafrost

Permafrost is defined as a condition—and not a material—in which a material remains below 0 °C for 2 or more years in succession (Muller 1943; van Everdingen 1998). Traditionally, this definition applies to rocks and buried ice as well as unconsolidated sediments, but it does not apply to glaciers, icings, or bodies of frozen surface water. Permafrost is synonymous with cryotic ground.

There are two general kinds of permafrost, including ice-cemented and dry-frozen. In the latter case, the material is below freezing but contains insufficient pore ice to be cemented. Ground ice is a general term referring to all types of ice contained in freezing and frozen ground.

2.4.2 Permafrost Boundaries and Distribution

Three zones of permafrost have been delineated, including continuous where permafrost covers 90 % or more of the surface, discontinuous where it covers between 50 and 90 %, sporadic where it covers 10–50 %, and isolated where it covers less than 10 % of the surface (Fig. 2.5). Permafrost covers 22.0×10^6 km^2, 24 % of the Earth's exposed land surface, but it likely covered up to 40 % during major glaciations (French 2007). About 99 % of the world's permafrost exists in the Northern Hemisphere (Fig. 2.5). Countries with the greatest areas of permafrost include the Russian Federation (11.0×10^6 km^2), Canada (6.0×10^6), China (2.1×10^6), and the USA (1.1×10^6 km^2) (Gruber 2012). In the Southern Hemisphere, permafrost occurs in the southern Andes (100,000 km^2) and ice-free areas of Antarctica (49,500 km^2) (Fig. 2.6). The distribution of permafrost is controlled by climate, elevation, latitude, snow cover, and other parameters.

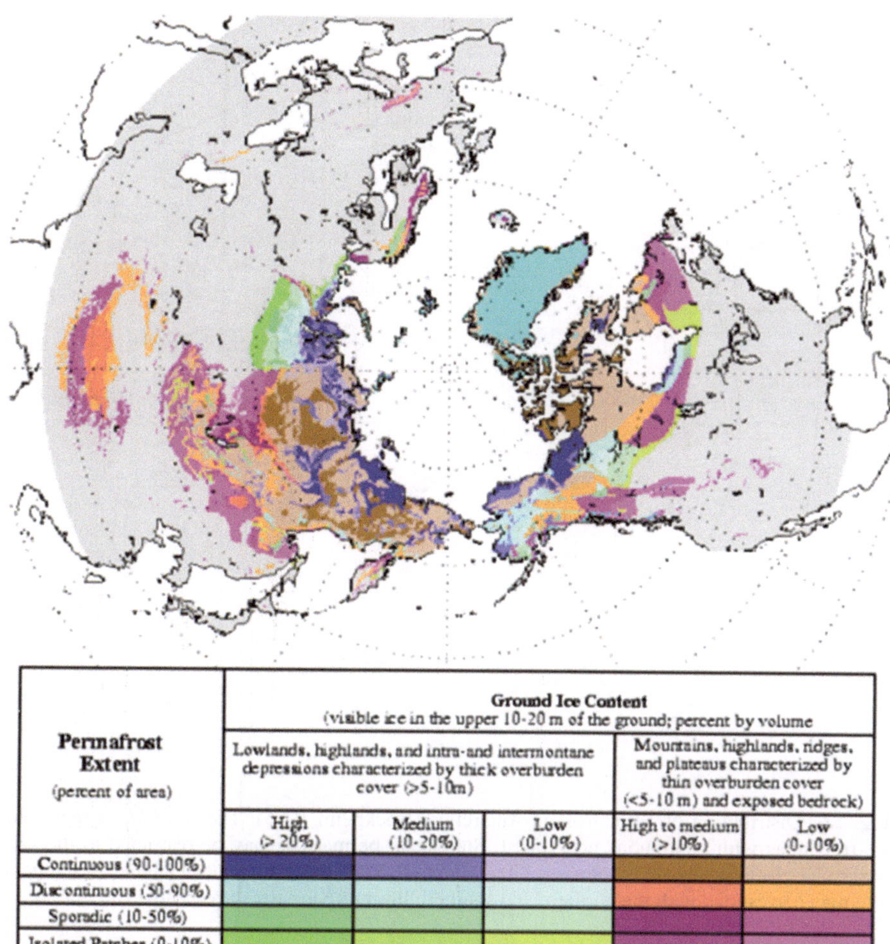

Fig. 2.5 Distribution of permafrost in the Northern Hemisphere (Brown et al. 1997)

2.4.3 Permafrost Properties

The thickness of permafrost ranges from less than a meter near its southernmost (Northern Hemisphere) or northernmost (Southern Hemisphere) boundary to over 1,500 m in northern Siberia. Buried ice in the Antarctic Dry Valleys may be 8 Ma in age (Marchant et al. 2002); ice-bonded permafrost in central Siberia is about 3 Ma in age and retains viable microorganisms (Gilichinsky et al. 1992). Permafrost temperatures are often reported as the Mean Annual Ground Temperature (MAGT), which is normally recorded at the 10-m or 20-m depth, and Temperature at the Top of Permafrost (TOPP). The temperature of permafrost ranges from near 0 °C in areas of sporadic permafrost to as low as −23 °C in the Antarctic Dry Valleys.

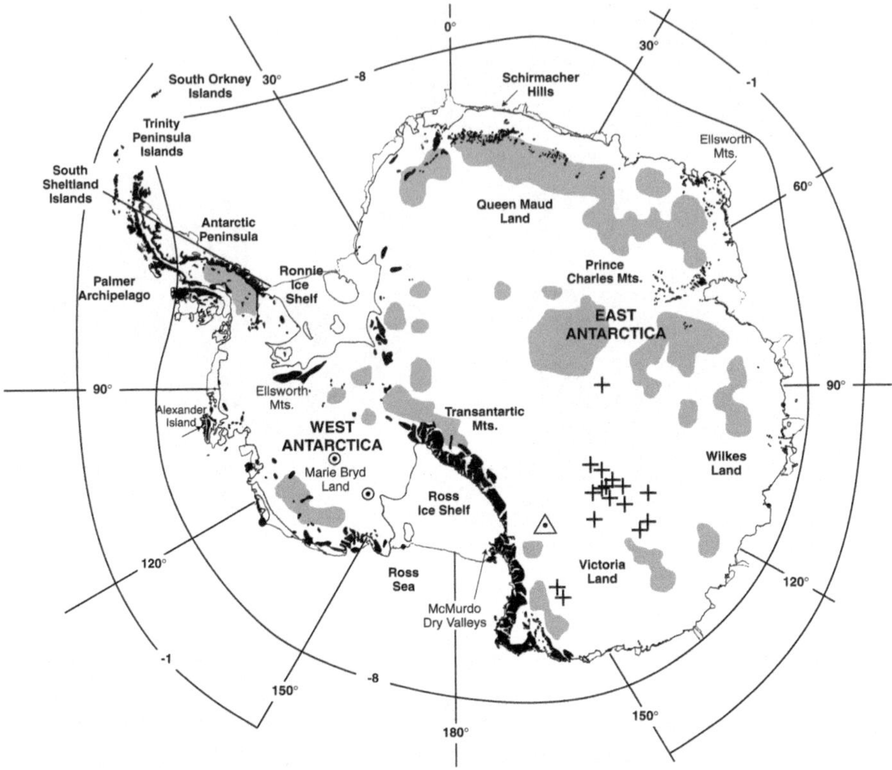

Fig. 2.6 Distribution of permafrost in Antarctica (Bockheim 1995). The *solid black* represents ice-free areas with continuous permafrost. Subglacial permafrost may be restricted to the *gray areas*. Subglacial lakes are by *crosses*. The –1 and –8 °C isotherms for mean annual air temperature may correspond with the northern and southern limit of permafrost, respectively

The "zero curtain" refers to the persistence of a nearly constant temperature close to the freezing point during annual freezing (and occasionally during thawing) of the active layer (Fig. 2.7). Despite that it is below freezing the upper part of permafrost is subject to temperature variations in response to air temperature variation. However, at a depth of 10–20 m, the permafrost temperature does not vary.

There is considerable variation in the ice content of permafrost. The ice content of permafrost is dependent on the texture of the material and the amount of "excess ice." Excess ice exists where the volume of ice in the ground exceeds the total pore volume of the material. The sandy dry-frozen materials in Antarctica may contain less than 5 % gravimetric water (Campbell and Claridge 2006). However, in cases where there is abundant excess ice, gravimetric water contents exceed 100 %.

The presence of segregation ice (ice in discrete lenses or veins formed by ice segregation) yields are variety of cryostructures, including basal, basal-layered, crust-like, layered, lens-type, massive-agglomerate, massive, massive-porous, reticulate and reticulate-blocky (French and Shur 2010). Some of these structural characteristics are depicted in permafrost cores from arctic Alaska (Fig. 2.8).

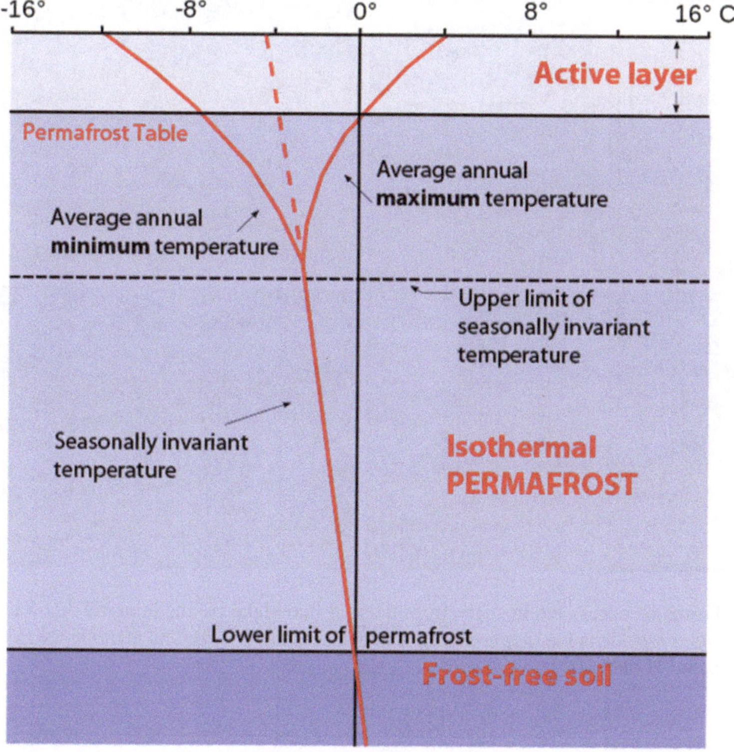

Fig. 2.7 Temperature profile for a region underlain by permafrost. The *red, dashed line* depicts the average temperature profile of the active layer and transient layer ("zero curtain"); the trumpet-shaped *solid lines* show the seasonal temperatures during freezing and thawing. The depth of the *horizontal dashed line* (zone of seasonally invariant temperature) is approximately 10–20 m (Source: Wikipedia)

Chemical properties of near-surface permafrost have been used to determine the age and origin of the water and organic matter (Kokelj and Burn 2005).

2.5 Summary

In this book the view is taken that the permafrost-affected soil is comprised of three parts: the active layer, the transition layer, and the permafrost layer. Soil formation is especially pronounced in the active layer but also occurs in the transition from warmer periods. The active layer varies from 0.1 m in high latitude environments to more than 10 m in low-latitude mountains. The active layer shows the maximum degree of cryoturbation, may contain high amount of organic C, often is dense from merging of freezing fronts, and may exhibit dilatancy in silt-rich soils.

Fig. 2.8 Forms of excess ice in cores from drained thaw-lake basins in arctic Alaska, including vein ice (*upper left*), lens ice (*upper right*), reticular ice (*lower left*), and atataxic ice (*lower right*) (Bockheim and Hinkel 2012)

The transition layer displays many of the same properties as the active layer, including high amounts of segregated ice, cryoturbation, and abundant soil organic C. Permafrost represents a condition in which a material remains at or below 0 °C for 2 or more years in succession. Permafrost may be over 1,500 m thick in central Siberia. Dry-frozen permafrost is a peculiar form of permafrost that occurs in hyper-arid regions of Antarctica. The nature of segregation ice in permafrost is important for determining its mechanism of formation.

References

Alekseev A, Alekseeva T, Ostroumov V, Siegert C, Gradusov B (2003) Mineral transformations in permafrost-affected soils, north Kolyma Lowland. Russia Soil Sci Soc Am J 67:596–605

Anisimov OA, Shiklomanov NI, Nelson FE (1997) Global warming and active-layer thickness: results from transient general circulation models. Global Planet Change 15:61–77

Bockheim JG (1995) Permafrost distribution in the Southern Circumpolar Region and its relation to the environment: a review and recommendations for further research. Permafr Periglac Process 6:27–45

Bockheim JG (2007) Importance of cryoturbation in redistributing organic carbon in permafrost-affected soils. Soil Sci Soc Am J 71:1335–1342

Bockheim JG, Hinkel KM (2005) Characteristics and significance of the transition zone in permafrost-affected soils of the Arctic Coastal Plain, Alaska. Arctic 58:406–417

Bockheim JG, Hinkel KM (2012) Accumulation of excess ground ice in an age-sequence of drained thermokarst lake basins. Arctic Alaska Permafr Periglac Process 23:231–236

Bockheim JG, Munroe JS (2014) Organic carbon pools and genesis of alpine soils with permafrost: a review. Arctic Antarct Alpine Res

Bockheim JG, Walker DA, Everett LR (1998) Soil carbon distribution in nonacidic and acidic tundra of arctic Alaska. In: Lal R, Kimble JM, Follett RF, Stewart BA (eds) Soil processes and the carbon cycle. CRC Press, Boca Raton, pp 143–155

Borden PW, Ping C-L, McCarthy PJ, Naidu S (2010) Clay mineralogy in arctic tundra soils, northern Alaska. Soil Sci Soc Am J 74:580–592

Brown J, Ferrians OJ Jr, Heginbottom JA, Melnikov ES (1997) Circum-arctic map of permafrost and ground-ice conditions. United States Geological Survey MAP CP-45, Denver, Colorado

Campbell IB, Claridge GGC (2006) Permafrost properties, patterns and processes in the Transantarctic Mountains region. Permafr Periglac Process 17:215–232

French HM (2007) The periglacial environment, 3rd edn. Longman, Essex, 478 p

French H, Shur Y (2010) The principles of cryostratigraphy. Earth-Sci Rev 101:190–206

Gilichinsky DA, Vorobyova EA, Erokhina LG, Fyordorov-Dayvdov DG, Chaikovskaya NR (1992) Long-term preservation of microbial ecosystems in permafrost. Adv Space Res 12(4):255–263

Gruber S (2012) Derivation and analysis of a high-resolution estimate of global permafrost zonation. The Cryosphere 6:221–233

Klene AE, Nelson FE, Shiklomanov NI, Hinkel KM (2001) The N-factor in natural landscapes: variability of air and soil-surface temperatures, Kuparuk River basin, Alaska, U.S.A. Arctic Antarct Alpine Res 33:140–148

Kokelj SV, Burn CR (2005) Geochemistry of the active layer and near-surface permafrost, Mackenzie delta region, Northwest Territories, Canada. Can J Earth Sci 42:37–48

Lewkowicz AG, Ednie M (2004) Probability mapping of mountain permafrost using the BTS method, Wolf Creek, Yukon Territory. Can Permafr Periglac Process 15:67–80

Mackay JR (1995) Active-layer changes (1968 to 1993) following the forest-tundra fire near Inuvik, NWT. Can Arctic Alpine Res 27:323–336

Marchant DR, Lewis AR, Phillips WM, Moore EJ, Souchez RA, Denton GH, Sugden DE, Potter N Jr, Landis GP (2002) Formation of patterned ground and sublimation till over Miocene glacier ice in Beacon Valley, southern Victoria Land. Antarct Geol Soc Am Bull 114:718–730

Muller SW (1943) Permafrost or permanently frozen ground and related engineering problems. US. Engineering Office, strategic engineering study of special report, 62. J.W. Edwards, Ann Arbor

Nelson FE, Outcalt SI (1987) A computational method for prediction and regionalization of permafrost. Arctic Alpine Res 19:279–288

Pecher K (1994) Hydrochemical analysis of spatial and temporal variations of solute composition in surface and subsurface waters of a high arctic catchment. Catena 21:305–327

Putkonen J (1998) Soil thermal properties and heat transfer processes near Ny-Ålesund, northwestern Spitsbergen. Svalbard Polar Res 17:165–179

Shur Y, Hinkel KM, Nelson FE (2005) The transient layer: implications for geocryology and climate-change science. Permafr Periglac Process 16(1):5–17

Tarnocai C, Bockheim JG (2011) Cryosolic soils of Canada: genesis, distribution, and classification. Can J Soil Sci 91(5):749–762. doi:10.4141/CJSS10020

van Everdingen R (ed) (1998) Multi-language glossary of permafrost and related ground-ice terms (revised May 2005). National Snow and Ice Data Center, Boulder

Vieira G, Bockheim J, Guglielmin M, Balks M, Abramov AA, Boelhouwers J, Cannone N, Ganzert L, Gilichinsky DA, Goryachkin S, López-Martínez J, Meiklejohn I, Raffi R, Ramos M, Schaefer C, Serrano E, Simas F, Sletten R, Wagner D (2010) Thermal state of permafrost and active-layer monitoring in the Antarctic: advances during the international polar year 2007–2009. Permafr Periglac Process 21:182–197

Zollinger B, Alewell C, Kneisel C, Meusburger K, Gärtner H, Brandová D, Ivy-Ochs S, Schmidt MWI, Egli M (2013) Effect of permafrost on the formation of soil organic carbon pools and their physical-chemical properties in the eastern Swiss Alps. Catena 110:70–85

Chapter 3
Description, Sampling, and Analysis of Cryosols

3.1 Describing Cryosols

Because of the underlying permafrost and cryoturbation, the approach for describing and sampling cryosols is different from that used for soils lacking permafrost. Schoeneberger et al. (2012) prepared a field book for describing and sampling soils. Guidelines for describing and sampling soils in Antarctica are given by Bockheim et al. for Antarctic soils (http://erth.waikato.ac.nz/antpas/pdf/Soil_Description_Manual-Draft_1.pdf).

The key attributes to be recorded for cryosols are given in Table 3.1. Properties that are somewhat unique to cryosols that should be described include patterned ground, cryoturbation features, the depth to the top and bottom of each horizon (for broken horizons), cryostructures, and the depth to and nature of permafrost. Patterned ground is discussed in more detail in Chap. 4. Cryoturbation is the dominant soil-forming process in cryosols (Chap. 5). A sketch illustrating the kinds of features that are indicative of cryoturbation is given in Fig. 3.1. Cryostructures are a key property of cryosols. Figure 3.2 illustrates several of the more important of these cryogenic structures. These are discussed further in Chap. 5.

A sketch and/or Polaroid image of the soil profile is important especially for cryoturbated soils with broken and irregular horizons (Fig. 3.3). By using two extendable measuring tapes, the horizons can be shown on a grid for later weighting to estimate overall horizon thicknesses.

The pedon concept is important in describing, sampling, and classifying cryosols. A pedon is considered to be the smallest volume of soil for descriptive and sampling purposes. The minimum horizontal area is set a 1 m^2 but it ranges to 10 m^2, depending on the variability of the soil. Figure 3.4a shows that the pedon selected encompasses the entire unsorted circle unit. In the case of large patterned ground units, such as low-centered polygons, two pedons are selected, one in the center of the polygon and the other in the contraction fissure (Fig. 3.4b).

J.G. Bockheim, *Cryopedology*, Progress in Soil Science,
DOI 10.1007/978-3-319-08485-5_3

Table 3.1 Components of a cryosol soil description

General
Pedon number
Collector – affiliation, contact address or email
Date
GPS location
Map grid reference
Photos of site, ground surface, soil profile, special soil features
Profile sketch
Site
Elevation
Topography-aspect, slope percent, complex, shape
Landform
Parent material – lithology and form
Patterned ground/cryoturbation
Climate regime
Vegetation/organisms
Weathering stage
Salt stage and type of salt
Presence/absence of ice cement – depth to ice cement
Depth to permafrost
Human impacted site – yes/no
Drainage class
Depth class
Soil classification – USDA, WRB, other
Soil taxon
Epipedon – kind and thickness
Diagnostic subsurface horizon – type and thickness
Soil moisture regime
Soil temperature regime
Soil description
Horizon name
Thickness of horizon
Depth to top and bottom of horizon
Percent of boulders, cobbles, gravel
Textural class
Munsell color by moisture status (dry, moist, wet)
Horizon boundaries
Structure
Consistence
Redox features
Concentrations
Ped surface forms
Roots
Pores
Other
Observation method

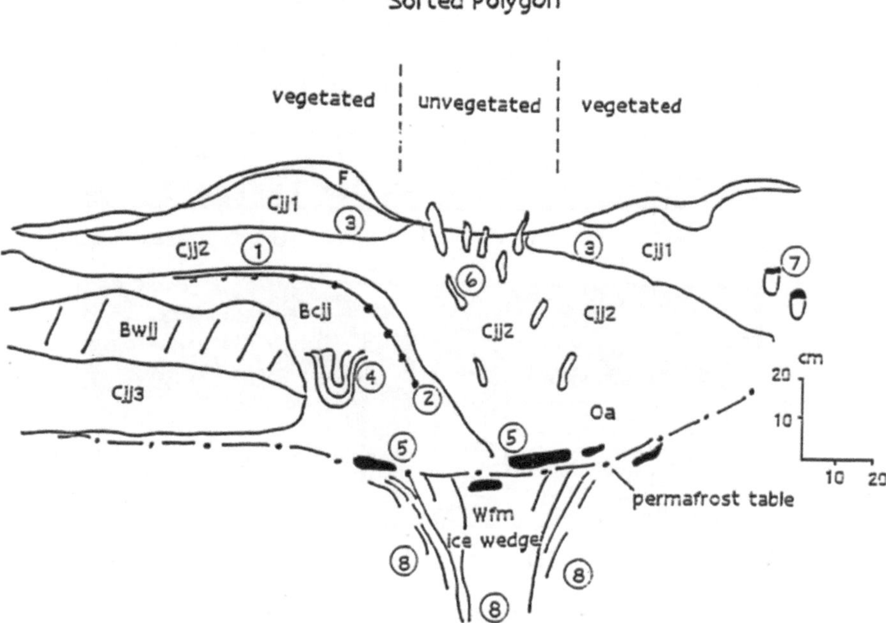

Fig. 3.1 Idealized sketch of an ice-wedge polygon showing types of cryoturbation features, including (*1*) irregular horizons, (*2*) deformation of textural bands, (*3*) broken horizons, (*4*) involutions and diapirs, (*5*) the accumulation of fibrous or partially decomposed organic matter on top of the permafrost table, (*6*) oriented stones, (*7*) silt caps from vertical sorting, and (*8*) upwarping of sediments adjacent to the ice wedge (Bockheim and Tarnocai 1998)

Table 3.2 compares soil horizon nomenclature for the US, Russian, Canadian, and WRB systems. There are several symbols and suffixes for soil horizon nomenclature that apply primarily to cryosols. The symbol "Wf" is used for describing ice wedges; the symbol "I" is used in the WRB. The suffixes "f" and "ff" are used for ice-cemented and dry-frozen permafrost, respectively. The suffix "jj" is used for horizons showing cryoturbation. As will be seen in Chap. 6, the use of these symbols is very important for classifying soils from soil descriptions.

The most common macroscopic soil features are due to cryoturbation and include irregular or broken horizons and incorporation of organic matter in lower horizons, especially along the top of the permafrost. Oriented stones and displacement of soil materials are common in cryosols. Freezing and thawing produce granular and platy structures in surface horizons and blocky, prismatic or massive structures in subsurface horizons. The massive structure is due to cryostatic pressure and desiccation that develop when the two freezing fronts, one from the surface and the other from the permafrost, merge during freeze back in the autumn. The perennially frozen layer commonly contains ground ice in the form of segregated ice crystals, vein ice, ice lenses and wedges, and thick ground ice.

The granular, platy, or blocky structures of the surface mineral horizons are also the result of cryopedogenic processes such as the freeze-thaw process and vein-ice

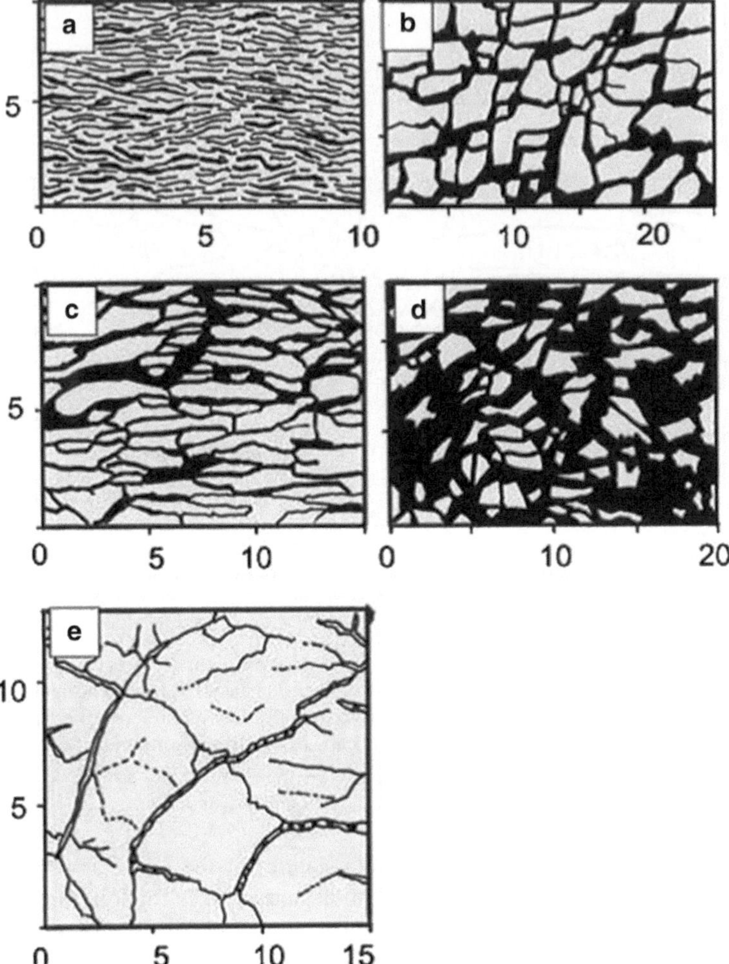

Fig. 3.2 Cryostructures present in cryosols, including (**a**) microlenticular, (**b**) reticulate, (**c**) renticular, (**d**) suspended (atataxitic), and (**e**) ice net. All scales are in cm and ice lenses in *black* (Ping et al. 2008)

formation (ice segregation process). The subsurface horizons often have massive structures and are associated with higher bulk densities, especially in fine-textured soils.

3.2 Sampling Cryosols

The preferred time for sampling cryosols is at the end of the summer, because the active layer is deepest at that time. Sampling earlier requires power tools to get to the base of the soil profile and a correction to be made for the end-of-season thickness

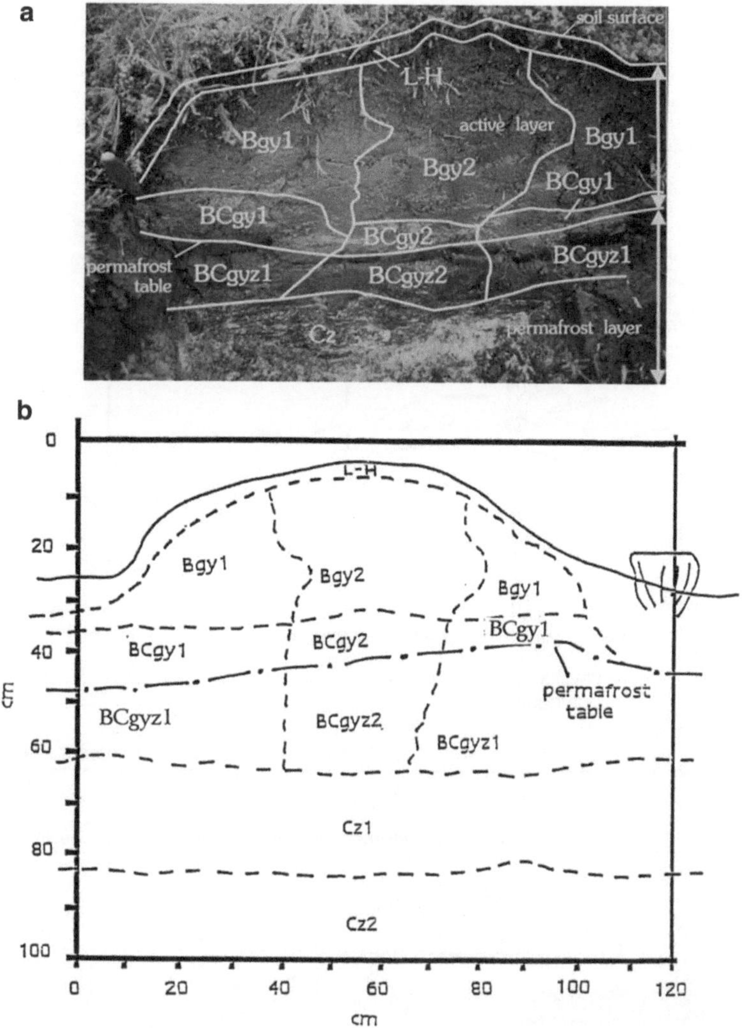

Fig. 3.3 A turf hummock in the Mackenzie delta, Canada including (**a**) a photograph of the profile with horizons delineated and (**b**) a sketch showing the horizons (Bockheim and Tarnocai 1998)

of the active layer. A description of the equipment used for coring, trenching, and handling frozen soil samples is given in Tarnocai (1993). The most common tool employed is a portable gasoline-powered concrete breaker (Fig. 3.5). The soil is excavated down to the ice-cement and the hammer is used to break up chunks that can be removed by hand to the desired depth. Carrying the breaker to remote sites and operating it for sustained periods of time can be very tedious. During the period of snow-cover, a gasoline-powered drill rig mounted on a sledge can be towed with a snow-machine to remote sites. The drill shown is equipped with a SIPRE core barrel for sampling to depths of 2 m or more in the frozen active layer and underlying permafrost (Fig. 3.6). However, these core barrels cannot be used in gravelly soils.

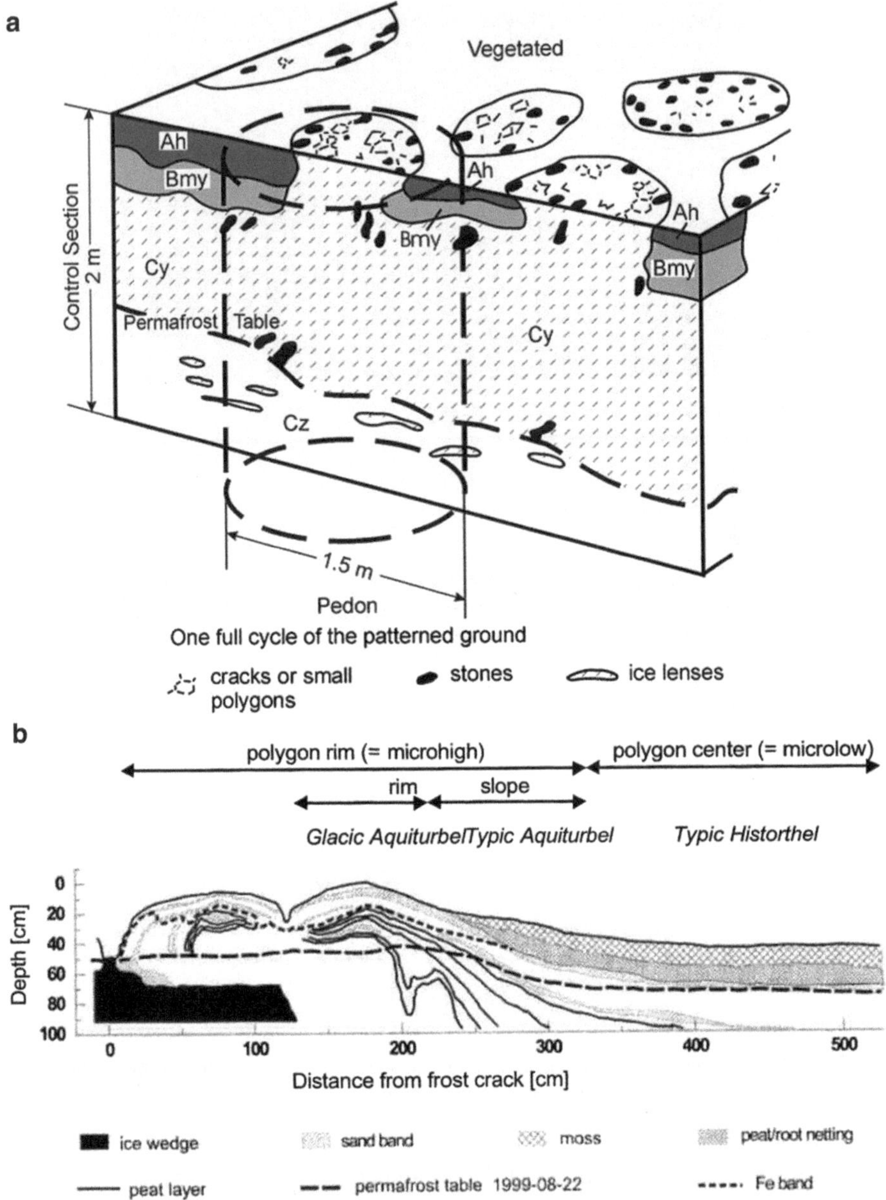

Fig. 3.4 Diagram showing: (**a**) The location and extent of the pedon in a cryoturbated cryosol in an unsorted circle type of patterned ground (Tarnocai and Bockheim 2011) and (**b**) A low-centered polygon in which a polygon would be selected to the far left over the ice wedge and to the far right under the center of the polygon (Ping et al. 2013)

Table 3.2 Soil horizon nomenclature for several taxonomic systems

Short description	Soil Taxonomy (Soil Survey Staff 2010)	Russian (Gerasimova et al. 2013)	Canadian (Soil Classification Working Group 1998)	WRB (IUSS Working Group 2006)
Master horizons				
Organic soil materials	O	O,T,TO,TE,TJ	L, F, H, O	L, O, H
Surface mineral horizon	A	A:AY,AU,AJ,AH,RU,H,AO	A	A
Bleached surface mineral horizon	E	EL,E	e	E
Illuvial horizon or weathered in situ	B	BT,BI,BHF/BM,BFM,BAN	B	B
Minimally weathered	C	BC,C	C	C
Hard bedrock	R	R	R	R
Root limit. Subsoil, human	M		–	–
Water	W		W	L, W
Suffixes				
Highly decomposed	a	h	h	a
Buried	b	[ ]	b	b
Concretions or nodules	c	nn,nc,fn	cc	c
Coprogenous earth	co		–	c
Physical root restriction	d	d,ad	–	d
Diatomaceous earth	di		–	d
Organic, intermed. decompose.	e	te,md	m	e
Frozen soil or water	f		z	f
Dry permafrost	ff	⊥	–	–
Strong gleying	g	g,q	g	l
Illuvial organic matter	h	BH,hi	h	h
Slightly decomposed organic	i	TJ, TO	f	i
Jarosite	j		–	j
Cryoturbation	jj	@	y	@

(continued)

Table 3.2 (continued)

Short description	Soil Taxonomy (Soil Survey Staff 2010)	Russian (Gerasimova et al. 2013)	Canadian (Soil Classification Working Group 1998)	WRB (IUSS Working Group 2006)
Secondary carbonates	k	mc,lc,ic,dc,nc	ca, k	k
Engulf secondary carbonates	kk		–	–
Cementation	m	(fn,ff)	c	m
Marl	ma	ML	–	M
Sodium	n	sn	n	n
Residual accum. sesquioxides	o		–	o
Tillage	p	agr+PU,P	p	p
Silica	q		–	q
Soft bedrock	r	CLM	–	s
Illuvial sesquioxides, organic matter	s	f	f	t
Slickensides	ss		ss	i
Silicate clay	t	?	t	t
Artifacts	u		–	u
Plinthite	v		–	v
Develop. color or structure	w		m	w
Fragipan	x		x	x
Gypsum	y	(cs)	–	y
Dominance gypsum	yy	cs	–	–
Salts more soluble than gypsum	z	s	sa	z
Strong reduction	–	G	–	r
Stagnant conditions	–	ox	–	g
Juvenile	–		–	j
Ice wedges	Wf		–	I
Vertisolization	–	v	u	
Shrink-swelling	–		v	

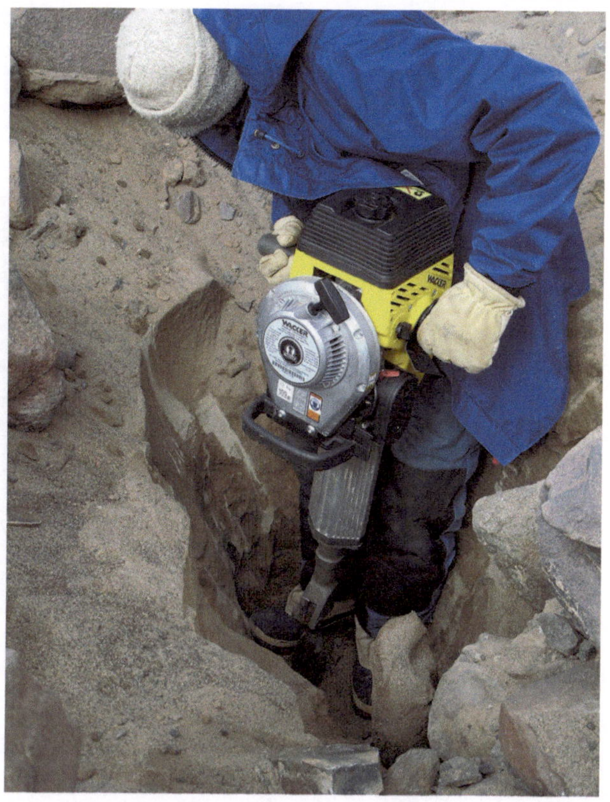

Fig. 3.5 Portable gasoline-powered hammer used to excavate frozen soil in Beacon Valley, Antarctica (Photo by M. Kurz)

Horizon sampling generally is employed for pedological studies; however, for specialized studies depth-sampling is used. For example, soil organic C budgets are often calculated for the 0–30 cm, 30–100 cm, and 100–300 cm depths (Hugelius et al. 2013). Frozen cores may be cut with a chop-saw into horizons or depth increments.

When sampling unfrozen cryosols, three kinds of samples are collected: bulk samples for routine analysis, minimally disturbed volumetric samples for bulk density and water retention measurement, and thin-section samples using special collection boxes for micromorphological analysis. Frozen cores require special handling. After they have been cut into horizons or sections, they should be placed in suitable containers that eventually will be allowed to thaw at room temperature. Once the samples have thawed, the excess water should be decanted and the volume noted for determining the excess water content. The samples are weighed and placed in a drying oven. The moisture content from this weighing represents that due to pores. We dry soil samples at 65 °C because drying at higher temperatures may cause combustion of organic matter. After a suitable time period, the samples are taken out of the oven and weighed; the pore-water content can then be calculated. The oven-dry bulk density can also be calculated. Removing and weighing coarse fragments may be important for some studies.

Fig. 3.6 Gasoline-powered drill rig with a SIPRE core barrel for sampling permafrost in arctic Alaska. The drill rig is carried on a sledge that is pulled by a snow machine (Photo by J. Kimble)

3.3 Laboratory Characterization of Cryosols

Cryosols, as with all soils, often require laboratory characterization, including physical, chemical, mineralogical, and micromorphological analyses. These methods are described in Burt (2004) and similar manuals.

3.3.1 Physical Analyses

Key physical properties that are analyzed include particle-size distribution (including further fractionation of clay, silt, or sand for mineralogical analysis) and bulk density. Water retention at 1,500 kPa may be measured and, when divided by the clay content, yields a rough measure of the potential water-holding capacity of the soil. The coefficient of linear extensibility (COLE) is a measure of shrink-swell capacity and is calculated from bulk density measured moist (at 33 kPa) and oven dry. Reynolds' dilatancy is an engineering test that measures the tendency of a compacted material to dilate (expand in volume) as it is sheared.

3.3.2 Chemical Analyses

Key chemical properties that are analyzed include pH (in water or a dilute salt suspension), organic C, total N, extractable P (Mehlich-1 is used for ornithogenic soils), total P, exchangeable bases (Ca, Mg, K, and Na), exchangeable acidity, extractable Al, cation-exchange capacity (CEC; pH 7 and/or 8.2), extractable Fe and Al (Na-pyrophosphate yields organic-bound forms; acid NH_4-oxalate yields organic and amorphous forms; and citrate-dithionite yields all "free" forms), and the optical density of the oxalate extract (ODOE). Effective (ECEC) represents the CEC from the sum of base cations and Al extracted with KCl. Salt-bearing soils should be analyzed for electrical conductivity, $CaCO_3$, and gypsum. Total elemental analysis is useful for evaluating the magnitude of weathering.

3.3.3 Mineralogical Analyses

Common mineralogical analyses include the use of optical techniques for the coarse silt (20–50 μm) and coarser fractions and X-ray diffraction for fine silt (5–2 μm), and clay (<2 μm) fractions.

3.3.4 Micromorphological Analyses

As with non-permafrost-affected soils (Bullock et al. 1985), microscopic examination of soils is important for evaluating soil-forming processes, particularly those related to cryogenesis. When viewed in thin sections, cryosols contain a variety of fabrics resulting from compaction (desiccation), displacement due to alignment, rotation, sorting, and inclusions, and pore formation (planar voids, cracks, vesicles) (see Chap. 5) (Fig. 3.7).

3.4 Summary

In addition to the usual properties described in low-latitude soils, properties that are unique to cryosols that should be described include patterned ground, cryoturbation features, the depth to the top and bottom of each horizon (for broken horizons), cryostructures, and the depth to and nature of permafrost. Sketches or field-ready photographs are recommended for diagramming soil horizons, cryoturbation, and other features. There are no universally accepted guidelines for delineating soil horizons; Table 3.2 compares horizon symbols used in the US, Russian, Canadian, and WRB systems.

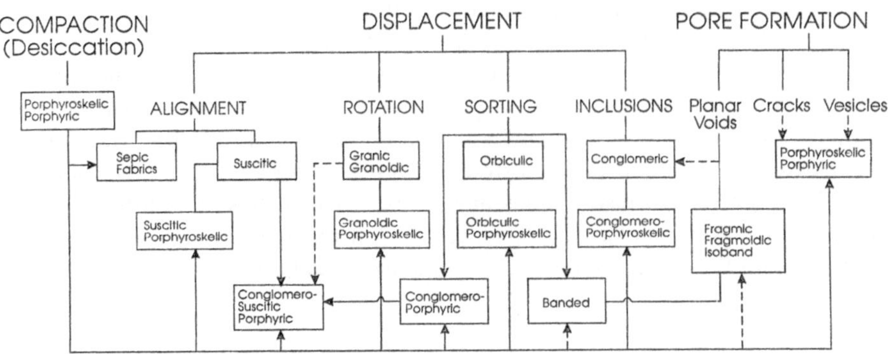

Fig. 3.7 Cryogenic processes evidenced from thin sections and micromorphology (Fox 1985)

The preferred time for sampling cryosols is at the end of the summer, because the active layer is deepest at that time. Sampling earlier requires power tools to get to the base of the soil profile and a correction to be made for the end-of-season thickness of the active layer. Three kinds of samples normally are collected from cryosols: bulk samples for routine analysis, minimally disturbed volumetric samples for bulk density and water retention, and thin-section samples for micromorphological analysis. Physical, chemical, mineralogical, and micromorphological analyses are normally performed for cryosols as with all soils.

References

Bockheim JG, Tarnocai C (1998) Recognition of cryoturbation for classifying permafrost-affected soils. Geoderma 81:281–293

Bullock P, Fedoroff N, Jongerius A, Stoops G, Tursina TE, Babel U (1985) Handbook for thin section description. WAINE Research Publication, Albrighton

Burt R (ed) (2004) Soil survey laboratory methods manual. Soil survey investigation report, no. 42, Ver. 4. U.S. Department of Agriculture, Natural Resource Conservation Service, Washington, DC

Fox CA (1985) Micromorphology of an Orthic Turbic cryosol-a permafrost soil. In: Bullock P, Murphy CP (eds) Soil micromorphology, vol 2, Soil genesis. Academic, Berkhamsted, pp 699–705

Gerasimova MI, Lebedeva II, Khitrov NB (2013) Soil horizon designation: state of the art, problems, and proposals. Eurasian Soil Sci 46:599–609

Hugelius G, Tarnocai C, Bockheim JG, Camill P, Elberling B, Grosse G, Harden JW, Johnson K, Jorgensen T, Koven CD, Kuhry P, Michaelson G, Mishra U, Palmtag J, Ping C-L, O'Donnell J, Schirrmeister L, Schuur EAG, Sheng Y, Smith LC, Strauss J, Yu Z (2013) Short communication: a new dataset for estimating organic carbon storage to 3 m depth soils of the northern circumpolar permafrost region. Earth Syst Sci Data Discuss 6:93–93

IUSS Working Group WRB (2006) World reference base for soil resources 2006, 2nd edn. World soil resources reports no. 103. FAO, Rome

Ping C-L, Michaelson GJ, Kimble JM, Romanovsky VE, Shur YL, Swanson DK, Walker DA (2008) Cryogenesis and soil formation along a bioclimate gradient in arctic North America. J Geophys Res 113:G0S12. doi:10.1029/2008JG000744

Ping C-L, Clark MH, Kimble JM, Michaelson GJ, Shur Y, Stiles CA (2013) Sampling protocols for permafrost-affected soils. Soil Horizons 54(1):13–19

Schoeneberger P, Wysocki DA, Behnam EE, Soil Survey Staff (2012) Field book for describing and sampling soils (vers. 3.0). National Soil Survey Center, National Resource Conservation Service, Lincoln

Soil Classification Working Group (1998) The Canadian system of soil classification, vol 1646, 3rd edn, Research Branch, Agricultural & Agri-Food Canada, publication. NRC Research Center, Ottawa

Soil Survey Staff (2010) Keys to soil taxonomy, 11th edn. U.S. Department of Agriculture, Natural Resource Conservation Service, Washington, DC

Tarnocai C (1993) Sampling frozen soils. In: Carter MR (ed) Soil sampling and methods of analysis. Taylor & Francis, Boca Raton, pp 755–767

Tarnocai C, Bockheim JG (2011) Cryosolic soils of Canada: genesis, distribution, and classification. Can J Soil Sci 91:741–762

Chapter 4
Factors of Cryosol Formation

4.1 Climate

4.1.1 Introduction

Other than the Aridisols in *Soil Taxonomy* (Soil Survey Staff 2010), the cryosol order is the only soil taxon in global and national systems that is based on climate. The developers of ST pointed out that the Aridisols, Gelisols, and suborders (Cry-, Gel-, Ud-, Ust-, Xer-) in most other orders are based on soil climate and not climate *per se* (Forbes 1986). Because of their opposition to the use of *soil* climate as a diagnostic soil property, developers of the WRB (IUSS WRB Working Group 2006) established the cryosol soil group based on the presence of a cryic horizon, "a perennially frozen soil horizon in mineral or organic materials" (p. 17), i.e., permafrost. However, in ST, the control section, or solum, constitutes the active layer and uppermost part of the transient layer and not the underlying permafrost (see Fig. 2.1). In ST, permafrost is viewed as a parent material and not soil material.

Regardless of these differences in philosophy, there is little doubt that climate, or soil climate, plays a critical, if not defining, role in the development of cryosols . In this chapter, the climate of regions supporting cryosols is discussed including temperature, precipitation and humidity, and wind—all of which are key to the development of cryosols.

In the widely used Köppen system, two types of polar climate are distinguished, ET, or tundra climate, and EF, or ice cap climate. However, this system does not recognize the considerable variation in mean annual air temperature (MAAT) and monthly temperatures in cryosol regions. French (2007) provided a useful classification of polar and alpine climates (Table 4.1). The scheme has five periglacial climate types that differ on the basis of latitude, elevation, and diurnal and seasonal patterns in air temperature, with the intensely cold, windy, and hyperarid Antarctic continent being a sixth type.

J.G. Bockheim, *Cryopedology*, Progress in Soil Science,
DOI 10.1007/978-3-319-08485-5_4

Table 4.1 Periglacial climates of the polar and alpine regions (French 2007)

1. *High Arctic climates* – in polar latitudes; extremely weak diurnal pattern, strong seasonal pattern. Small daily and large annual temperature range. Examples: Spitsbergen (Green Harbour, 78°N); Canadian Arctic (Sachs Harbour (Ikaahuk), 72°N).
2. *Continental climates* – in sub-arctic latitudes; weak diurnal pattern, strong seasonal pattern. Extreme annual temperature range. Examples: Central Siberia (Yakutsk, 62°N); Interior Alaska and Yukon Territory (Fairbanks, 65°N; Dawson City, 64°N).
3. *Alpine climates* – in middle latitudes in mountain environments; well-developed diurnal and seasonal patterns. Examples: Colorado Front Range, USA (Niwot Ridge, 40°N); European Alps (Sonnblick, 47°N).
4. *Qinghai-Xizang (Tibet) plateau* – a high-elevation, low-latitude mountain environment. Well-developed diurnal and seasonal patterns. Above-normal insolation due to elevations of 4,200–4,800 m a.s.l. Example: Fenghuo Shan (34°N).
5. *Climates of low annual temperature range* – two types of azonal locations: (a) island climates in sub-arctic latitudes. Examples: Jan Mayen (71°N), South Georgia (54°S). (b) Mountain climates in low latitudes. Examples: Andean summits, Mont Blanc Station, and El Misti, Peru (16°S); Mauna Kau, Hawaii (20°N).
6. *Antarctic climate* – intense cold, windiness, and aridity of the ice-free areas of the Antarctic continent.

4.1.2 Temperature

The MAAT ranges from +1 °C in areas of sporadic or isolated permafrost to −20 °C or colder in the northern Canadian Arctic Archipelago and ice-free areas of the interior Antarctic mountains (Table 4.2). There is considerable variation in the amplitude of mean monthly temperatures in cryosol regions. In cryosol areas with a maritime climate, as on Svalbard, Norway or King George Island, Antarctica, the variation from the coldest to the warmest month may be only 17–21 °C. In contrast the annual monthly variation in cryosol areas with an ultra-continental climate, such as Yakutsk, Siberia, may be as high as 55°. The temperature of the warmest month is strongly dependent on latitude and elevation. The mean July temperature varies from 2 to 6 °C in the High Arctic, from 5 to 12 °C in the Mid-Arctic, and from 6 to 19 °C in the Low Arctic. In high-mountain areas with permafrost, the mean July temperature ranges between 4 and 11 °C.

4.1.3 Precipitation and Humidity

There is also considerable variation in mean annual and monthly variations in precipitation and in the form of precipitation in cryosol areas. In Antarctica the MAP varies from <10 mm in the McMurdo Dry Valleys to over 1,000 mm along the western Antarctic Peninsula (Table 4.2).

However, many areas in the Arctic and the high mountains of central Asia receive between 200 and 400 mm year^{-1} of water equivalent (w.e.) precipitation. Many high-mountain locations receive between 1,400 and 2,000 mm of precipitation

Table 4.2 Generalized climate data for Arctic, Antarctic and Alpine locations

Location	Latitude (°)	Longitude (°)	Elevation (m)	MAAT (°C)	Mean July temp. (°C)[a]	MAP (mm/year)	Mean ann. snowfall (m)	Permafrost zone[b]
Arctic								
Russia								
Arctic Archipelagos	71–81N	52–180E	5–100	−1.5 to −10	1.0–6.0	125–300	1–3	C
Northeastern Eurasia	60–73N	140E–170W	5–500	−4.1 to −16	4.0–12	100–600	1–4	C
European North	67–70N	30–65E	5–200	−0.8 to −7.6	5.8–12	270–800	1–4	C, D
Western Siberia	62–73N	65–100E	5–500	−3.4 to −11	5.7–17	115–400	1–3	C
Central Siberia	54–78N	80–135E	5–500	−0.5 to −16	11–19	150–1,000	<1–8	C, D
Canada								
Low-Arctic	58–69N	141–63W	5–200	−8.2 to −14	8.9–11	150–200	1–2	C
Mid-Arctic	63–75N	125–65W	5–500	−15 to −16	4.5–4.9	100–250	1–2	C
High-Arctic	67–84N	125–67W	5-500	−17 to −19	2.0–4.0	100–175	0.5–1	C
Alaska								
Arctic Coastal Plain	70–72N	145–163W	5–100	−10 to −13	4.9–7.6	125–150	<1	C
Arctic Foothills	68–70N	142–163W	100–500	−8.3 to −8.8		140–270	1	C
Western Alaska lowlands	64–67N	160–167W	100–500	−0.5 to −4.0	11–15	250–500	1	D
Interior Alaska	64–66N	140–167W	100–500	−2.5 to −6.0	15–17	230–340	1.5	D
Greenland								
Southern	60–68N	20–55W	5–500	−1.3 to −2.0	6.4–6.7	775–980	3–5	C
Central	68–76 N	19–75W	5–500	−4.4 to −7.2	5.2	250–275	1–3	C
Northern	76–83N	12–67W	5–500	−12 to −17	3.7	140–200	1–3	C
Antarctic								
Western Antarctic islands, peninsula	60–65S	45–64W	5–90	−1.7 to −3.4	−0.1 to 2.0	400–800	3–6	D
East Antarctic coast	66–69S	40–100E	8–42	−9.2 to −11.4	0 to −2.0	200–380	1–2	C
Transantarctic Mtns.	75–78S	162–167E	24–150	−18 to −20	0.8 to −3.6	5–200	<1	C

(continued)

Table 4.2 (continued)

Location	Latitude (°)	Longitude (°)	Elevation (m)	MAAT (°C)	Mean July temp. (°C)[a]	MAP (mm/year)	Mean ann. snowfall (m)	Permafrost zone[b]
Alpine								
Western Cordillera, USA	37–49N	105–115W	>3,500	1.1 to –3.3		635–1,400	4–12	I
	60–70N							S, D
Western Cordillera, Canada	51–60N	115–140W	>2,180–>3,000	–1.0 to –1.7		280–1,400	2–12	C, D
Brooks Range, USA	68–69N	143–161W	>500	–5 to –10		180–390	1–2	C
Appalachian Mountains, USA	42–45N	71W	>1,200-1,800	1.4 to –3.0	2.0–6.0	1,800	6–8	I
Andes Mountains, South America	20–57S	70–78W	>1,500–>5,000	1.4 to –4	8	300–900	1–7	S, D
Fennoscandian mountains	63–70N	14–30E	>1,400–>1,800	–3.9 to –6.1	1.4–11	450–1,200	4–12	C, D, S
Iceland	64–66N	14–23E	>800–>1,000	–1.8 to –4.5	10–11	600–1,500	2–3	C, I
Svalbard	77–78N	11–25E	>30	–3.8 to –4.4	4.4	200–800	1–3	C, I
European Alps	44–47N	6–15E	>2,400–>3,000	–3.0 to –5.5	6	1,100–2,400	4–15	C, D, S
Pyrenees, Europe	42N	1W–2E	>2,700	0 to –0.7		1,600–2,000	12–16	C, D, S
Carpathians, Europe	44–48N	24–27E	>2,100	–2.6 to –4.0		850–2,000	2–9	
Urals, Russia	50–60N	58–60E	>550–>1800	–3 to –7		570–1,000	2–4	
Caucasus, Russia	41–43N	40–48E	>2,800–>3,000	–3 to –11		500–2,500	3–10	
Qinghai-Tibet Plateau	30–40N	70–105E	>4,600–>5,000	–1.1 to –7.1		300–500	2–4	C, D, S
Altai Mtns., central Asia	45–53N	87–97E	>2,400–>3,600	–1.0 to –4.0	10–11	360	1–5	C, D, S
Tien Shan Mtns., Kyrgyzstan, China	38–44N	69–95E	>2,700–>3,000	1.0 to –3.5	10–11	700–2,000	3–16	C, D, S
Sayan-Yablonoi Mtns., Siberia	50–52N	92–120E	>1,750	–3.1		800	3–10	
Japanese Alps	44N	143E	>1,600–>2,000	–2.8 to –3.8	10	2,200	3–11	I

[a]Mean Dec. or Jan. temperature for Southern Hemisphere

[b]Permafrost zonation: *C* Continuous, 90–100 %, *D* Discontinuous, 50–90 %, *S* Sporadic, 10–50 %, *I* Isolated, <10 %

annually. Some cryosol areas receive most of their precipitation during the summer months as rain, such as Yakutask, Siberia; other areas, such as high mountains with permafrost, receive most of their precipitation during the winter as snow.

The polar regions vary markedly in terms of the magnitude and seasonal variation in relative humidity. Areas close to the Arctic Ocean tend to be humid (70–90 %). In contrast the dry valleys of Antarctica have a humidity of less than 20 %, especially during the winter months when katabatic winds pour off the polar plateau. A major concern in the arctic is an apparent rise in humidity, which is attributed to melting of sea ice and warming of the Arctic Ocean.

4.1.4 Wind

As with most of the climate parameters, there is considerable variation in diurnal, seasonal, and mean annual wind velocities in areas with cryosols. In the Canadian Arctic Archipelago, mean annual wind speeds range between 4 and 6 m s^{-1}; in contrast, many alpine areas and continental Antarctica have exceptionally high wind speeds. For example, the mean annual wind velocity on Mount Washington, New Hampshire is 16 m s^{-1}. Wind is normally considered an important soil-forming factor in alpine and Antarctic environments.

4.2 Biota

Although climate is the most important soil-forming factor in the cryosol region, the biota resulting from climate, including plants and animals, are also important.

4.2.1 Vegetation

The circumarctic is commonly subdivided by latitude, permafrost continuity, and vegetation into the subarctic, Low Arctic, Mid-Arctic, and High Arctic regions (Fig. 4.1). The subarctic or taiga contains primarily open boreal forest and is underlain by sporadic or isolated permafrost. The low arctic contains low forms of vegetation, primarily peaty and mineral graminoid tundra, and is underlain by discontinuous permafrost. The mid-arctic also has tundra vegetation and is underlain by discontinuous permafrost. The high arctic or polar desert contains cryptogam, herb barrens and mountain complexes and prostrate and semi-prostrate dwarf-shrub, herb tundra and is underlain by continuous permafrost. The vegetation types within each region are illustrated in Fig. 4.2. About 26 % of the vegetated area is erect shrub land, 18 % peaty graminoid tundra, 13 % mountain

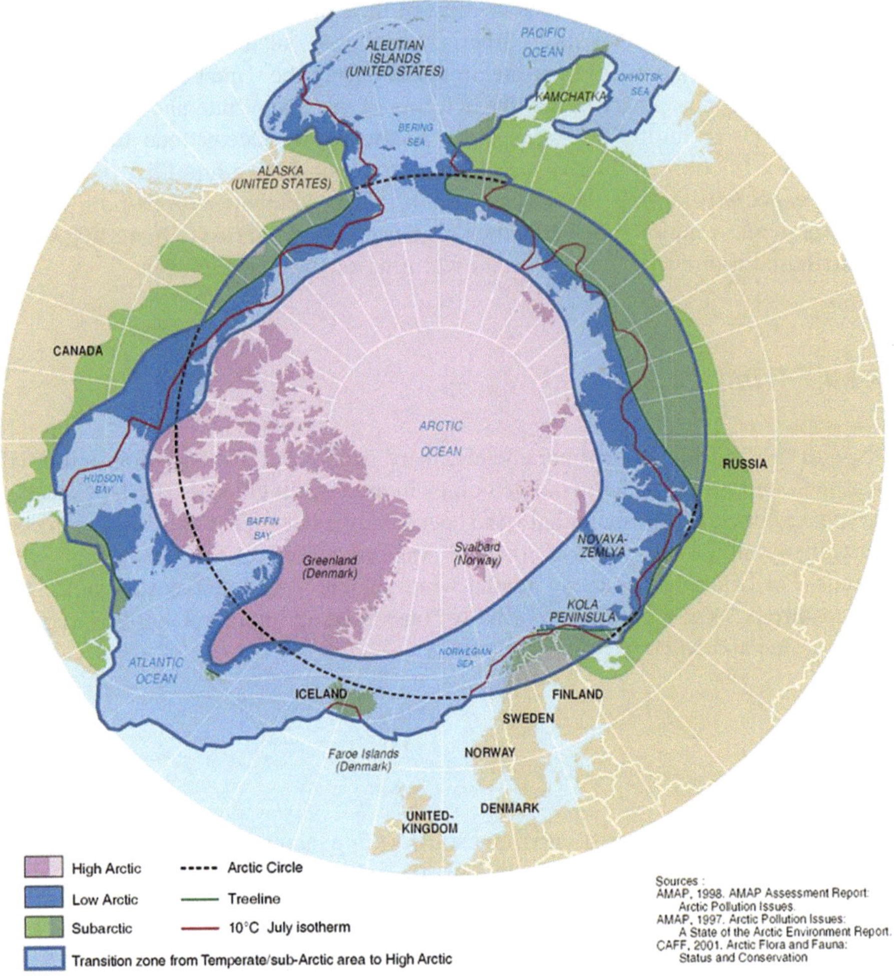

Fig. 4.1 Physiographic-botanic zones of the circumarctic region (Image by AMAP)

complexes, 12 % barrens, 11 % mineral graminoid tundra, 11 % prostrate-shrub tundra, and 7 % wetlands (Walker et al. 2005).

Only 0.35 % of Antarctica is ice-free, and the ice-free regions are distributed like a "no" or "don't" sign along the coast and across the continent as the Transantarctic Mountains (see Fig. 2.6). Longton (1979) reviewed the plant community ecology within the "Antarctic Botanical Zone." The vegetation is formed largely of mosses, lichens, and algae, but two native species of flowering plants (*Deschampsia Antarctica* and *Colobanthus quitensis*) occur along the western Antarctic Peninsula from 60 to 67°S. He was able to recognize eight physiognomically distinct subformations of cryptogam tundra (Fig. 4.3).

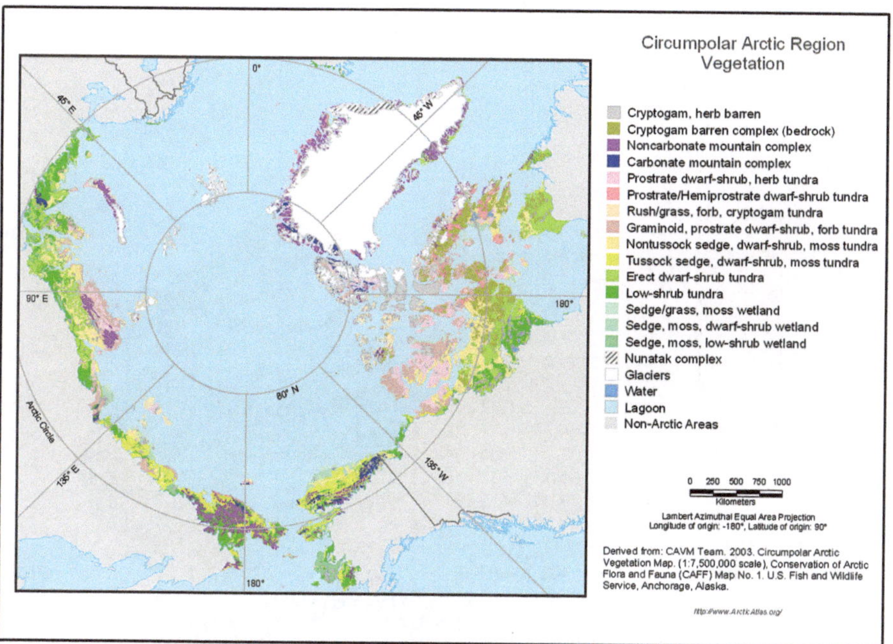

Fig. 4.2 Circumpolar arctic region vegetation (Source: D.A. Walker)

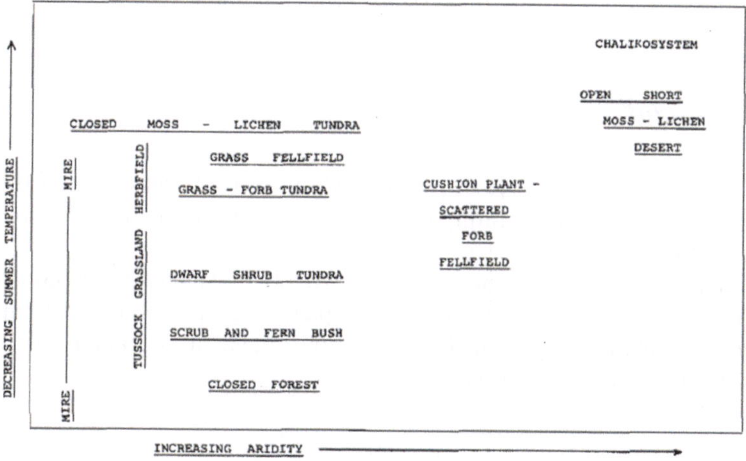

Fig. 4.3 A two-dimensional classification of southern circumpolar vegetation (Longton 1979)

The Qinghai-Tibet Plateau has been referred to as the "Third Pole." The plateau has an average elevation of 4,000 m, covers an area of 2.5 million km², and has an estimated 1.3 million km² of permafrost. The plant communities on the QTP are commonly subdivided into alpine steppe, alpine meadow, and alpine cushion plants (Wu et al. 2012) (Fig. 4.4).

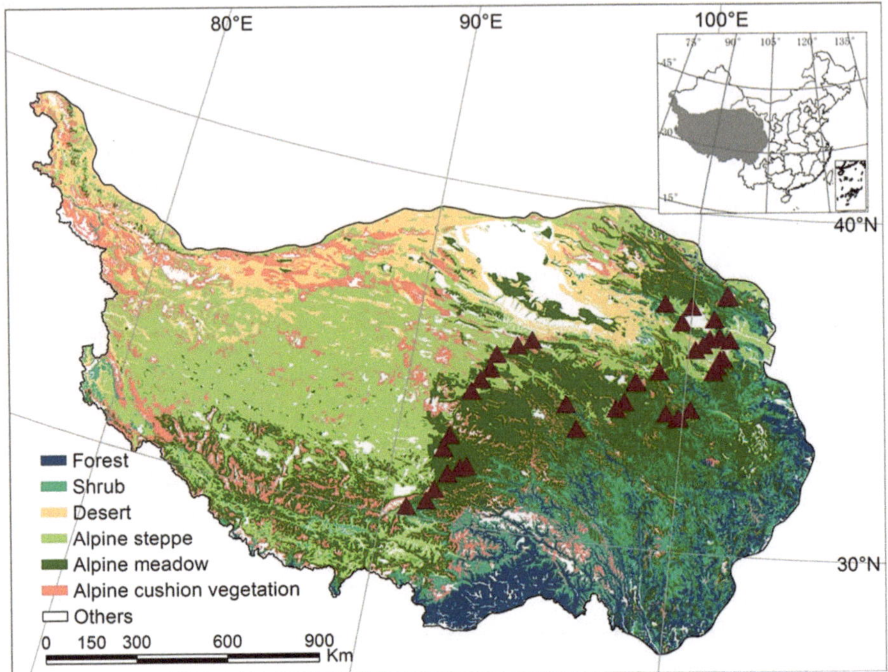

Fig. 4.4 Vegetation of the Qinghai-Tibet Plateau (Source: Northwest Institute of Plateau Biology, CAS)

4.2.2 *Animals*

Animals play a dominant role in pedogenesis in coastal areas of both the arctic and Antarctic, including migrating birds, penguins (Antarctica), seals, microtines (arctic), and other organisms.

Ornithogenic soils have been studied extensively in Antarctica and contain high levels of organic C and N, and calcium phosphates (Tatur 1989). Contributions of these constituents are important for the establishment of vegetation in the polar regions (Convey and Smith 2006).

4.2.3 *Microorganisms*

Microorganisms are ubiquitous in cryosols and include gram-positive and gram-negative bacteria, archaea, yeasts, filamentous fungi, and cyanobacteria (Margesin and Miteva 2010).

There are significant numbers of viable ancient microorganisms within permafrost. They have been isolated in both polar regions from cores up to 400 m deep and ground temperatures of −27 °C (Gilichinsky et al. 2008). The age of the cells dates back to ~3 million year in the arctic and ~5 million year in Antarctica. They are the only life forms known to have retained viability over geologic time.

4.2.4 Humans

Humans have played a major role in soil development over the past 100–150 years. In the arctic, mining activities and extraction of petroleum products have impacted soils. In the Antarctic there are some 72 scientific bases that have left a small but detectable influence on the terrestrial and aquatic ecosystems. The management of cryosols at all three "poles" is considered in Chap. 13.

4.3 Patterned Ground

4.3.1 Patterned Ground Forms

Patterned ground is a general term for any ground surface exhibiting a discernibly ordered, more or less symmetrical, morphological pattern of ground and, where present, vegetation. Some patterned ground features are not confined to permafrost regions, but they are best developed in regions of present or past intensive frost action.

A descriptive classification of patterned ground includes sorted and poorly sorted or unsorted circles, nets, polygons, steps, stripes, and solifluction features. Several of the sorted forms are illustrated in Fig. 4.5. In permafrost regions, the most ubiquitous macro-form is the ice-wedge polygon, a common micro-form is the unsorted circle called frost boils, and earth hummocks (Fig. 4.6). Patterned ground also occurs in peat lands in the form of strangmoor and frost mounds (Fig. 4.6). Detailed discussions of patterned ground features are given in Washburn (1956, 1980).

4.3.1.1 Ice-Wedge Polygons

Ice-wedge polygons are formed by thermal cracking of the underlying permafrost during winter freezing. The cracks project upward to the surface. During thawing of the active layer in spring, water flows down into the crack. The cracks gradually build up horizontally, and horizontal compression produces upturned of the frozen sediments by plastic deformation. Eventually the cracks may become several meters

Fig. 4.5 Photos of sorted patterned ground features in Iceland, including (**a**) a sorted polygon, (**b**) a sorted circle, (**c**) a large active polygon, (**d**) small active polygons, (**e**) miniature active circles, and (**f**) large stone circles (Feuillet et al. 2012)

wide and several meters deep. This process is depicted in Fig. 4.7. Ice-wedges cover as much as 23 % of the surface in drained thaw-lakes of arctic Alaska (Brown 1967). Ice-wedge polygons create considerable mesosoil variability. In Fig. 4.8 the SOC varied by nearly threefold from the ice wedge to the center of high-centered and flat-centered polygons in arctic Alaska.

4.3.1.2 Frost Boils

Mud boils (also known as frost boils and mud circles) are upwellings of silt and clay from frost heave and cryoturbation. They are typically 1–3 m in diameter on a bare soil surface, are dominantly circular, and unlike stone circles lack a stone border (Daanen et al. 2008; Michaelson et al 2008; Ping et al. 2008). The mechanism for the formation of frost boils is given in Fig. 4.9.

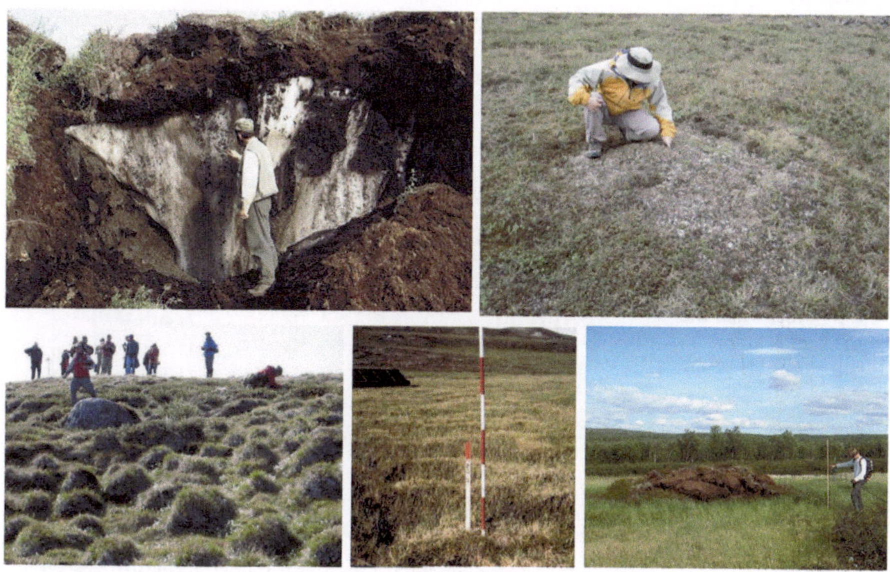

Fig. 4.6 Examples of non-sorted patterned ground: ice-wedge polygon in the Canadian Arctic (*upper left*; photo by Agriculture and Agric-Food Canada; a frost boil in the Alaskan Arctic Foothills; photo by J. Bockheim); earth or turf hummocks in Greenland (*lower left*; photo by Philipp Gaugler; http://www.tuhlig.de); strangmoor in the Alaskan Arctic Foothills (*lower center*; photo by D.A. Walker); and a palsa near Tromso, Norway (*lower right*; photo by sciencenordic.com)

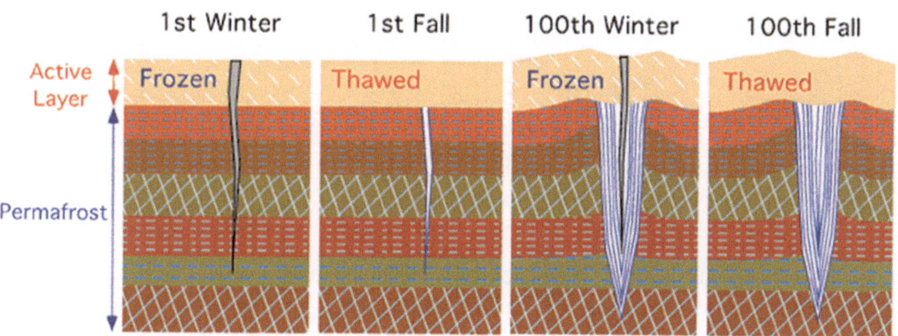

Fig. 4.7 Formation of ice-wedge polygons (Pidwirny 2006)

4.3.1.3 Earth Hummocks

Earth hummocks were studied in detail by Tarnocai and Zoltai (1978) in the Canadian arctic and subarctic. The hummocks have an average diameter of 80–160 cm and an average height of 40–60 cm. They develop on mineral materials and have abundant ice. The development of earth hummocks is controlled by soil

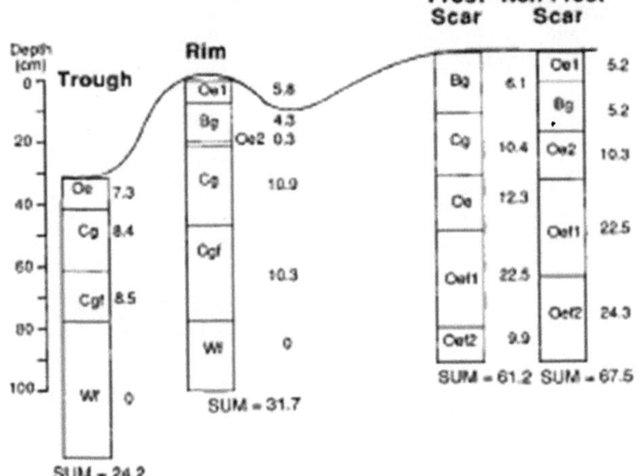

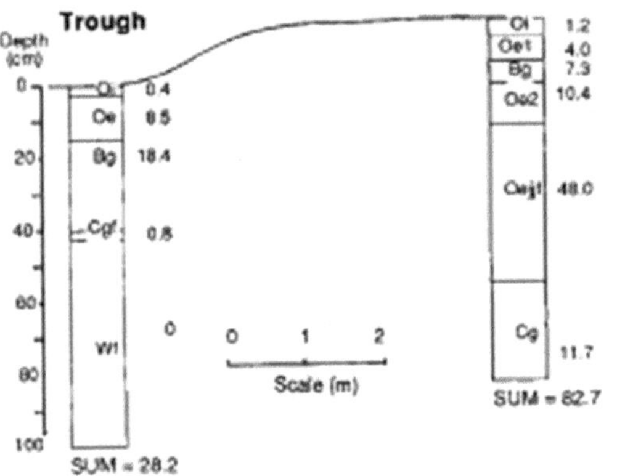

Fig. 4.8 Soil organic C within (**a**) a high-centered polygon in the Barrow, Alaska Environmental Observatory and (**b**) a flat-centered polygon at the former International Biological Program Tundra Biome site, Barrow, AK. Values to the *right* of each horizon are reported in kg C m^{-2} $year^{-1}$ and the profile sum is kg m^{-3} (Bockheim et al. 1999)

texture, soil moisture, and soil temperature. Most earth hummocks in the Mackenzie River valley have formed during the last 5,000 year.

4.3.1.4 Frost Mounds

Frost mounds are peaty permafrost mounds with a core of alternating layers of segregated ice and peat or mineral soil material. Palsas, a type of these features, have been studied extensively by Seppälä (2006). Frost mounds are typically between 1 and 10 m in height and a few meters to 100 m in diameter. Most, but not all, frost mounds occur in the discontinuous permafrost zone. Frost mounds form from permafrost aggradation at an active-layer-permafrost contact zone or

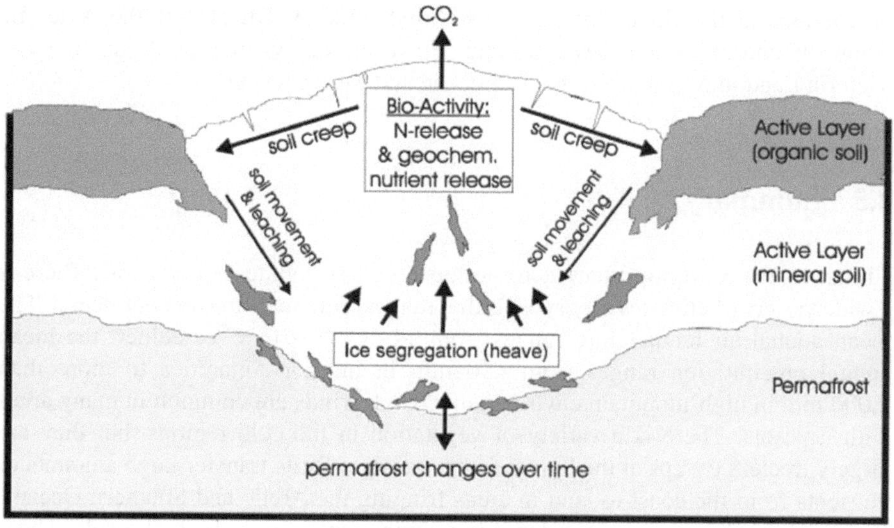

Fig. 4.9 Conceptual diagram of biogeochemical cycling within frost boils (Walker et al. 2004)

from intrusion of ice. Palsa contain ice lenses which grow during the autumn when ground frost penetrates the peat. Wind keeps the top of palsa nearly snowless so that it is more susceptible to frost. The ice lenses also raise the underlying mineral soil.

4.4 Parent Materials and Time

As with most soil orders, cryosols have formed in a variety of parent materials. In the circumarctic, the most common parent materials are eolian (loess and sand dunes), alluvium, lacustrine deposits, peat, colluvium, gelifluction materials, and beach sediments. In western Alaska and the Russian Far East, ash from the circum-pacific volcanic rift is common. In Antarctica the most common parent material is till, but glaciofluvial and outwash materials, colluvium, and gelifluction materials also occur. Residual soils occur on nuntaks in high-mountain areas of Antarctica, including the Sør Rondane, the Prince Charles, Transantarctic, Ellsworth, Pensacola, Shackleton Mountains and mountains along the spine of the Antarctic Peninsula. In alpine environments, the most common parent materials are till, colluvium, residuum, and mass wasting deposits.

Several large areas in the arctic were not glaciated during the Last Glacial Maximum (LGM) and soils in these areas may be very old. In the Arctic Coastal Plain, the thaw-lake cycle keeps landforms to an age of <5,500 year. Despite pervasive cryoturbation, soil properties still could be used to assess the relative age of soils in the Low-Arctic (Munroe and Bockheim 2001). Soils tend to be of LGM age or younger in the South Shetland and South Orkney Islands off the Antarctic Peninsula and in coastal East Antarctic. However, valleys of central and southern Victoria Land may contain soils of Pliocene and Miocene age.

4.5 Summary

Climate is the most important factor influencing development of cryosols. There is a wide variety of climates for cryosols that ranges from maritime to continental. The mean annual air temperature ranges from +1 °C to −20 °C or colder; the mean annual precipitation ranges from <10 mm in interior Antarctica to more than 2,000 mm in high-mountain environments. High winds are common in many areas with cryosols. There is a variety of vegetation in the cold regions, but they are largely treeless except in the boreal forest or taiga. Birds transfer large amounts of nutrients from the coast to land in areas fringing the Arctic and Southern Oceans. Patterned ground is a common landform component in areas with cryosols. These features may be sorted in circles, nets and stripes, or they may be unsorted in the form of ice-wedge polygons, frost boils, and earth hummocks. The most common parent materials are eolian (loess and sand dunes), alluvium, lacustrine, peat, colluvium, gelifluction materials, and beach sediments. These materials may be of recent age or may extend back as far as the Miocene.

References

Bockheim JG, Everett LR, Hinkel KM, Nelson FE, Brown J (1999) Soil organic carbon storage and distribution in arctic tundra, Barrow, Alaska. Soil Sci Soc Am J 63:934–940

Brown J (1967) Tundra soils formed over ice wedges, northern Alaska. Soil Sci Soc Am Proc 31:686–691

Convey P, Smith RIL (2006) Responses of terrestrial Antarctic ecosystems to climate change. Plant Ecol 182:1–10

Daanen RP, Misra D, Epstein H, Walker D, Romanovsky V (2008) Simulating unsorted circle development in arctic tundra ecosystems. J Geophys Res 113:G03S06. doi:10.1029/2008JG000682

Feuillet T, Mercier D, Decaulne A, Cossart E (2012) Classification of sorted patterned ground areas based on their environmental characteristics (Skagafjördur, northern Iceland). Geomorphology 139–140:577–587

Forbes TR (1986) The Guy Smith interviews: ratinale for concepts in soil taxonomy, SMSS technology monograph 11. Soil Management Support Services, Washington, DC

French HM (2007) The periglacial environment, 3rd edn. Wiley, New York

Gilichinsky D, Vishnivetskaya T, Petrova M, Spirina E, Mamykin V, Rivkina E (2008) Bacteria in permafrost. In: Margesin R et al (eds) Psychrophiles: from Biodiversity to Biotechnology. Springer, Berlin, pp 83–102

IUSS Working Group WRB (2006) World reference base for soil resources 2006, 2nd edn, World soil resource report no. 103. FAO, Rome

Longton RE (1979) Vegetation ecology and classification in the Antarctic Zone. Can J Bot 57:2264–2278

Margesin R, Miteva V (2010) Diversity and ecology of psychrophilic microorganisms. Res Microbiol 162:346–361

Michaelson GJ, Ping CL, Epstein H, Kimble JM, Walker DA (2008) Soils and frost boil ecosystems across the North American Arctic Transect. J Geophys Res 113:G0S11. doi:10.1029/2007JG000672

Munroe JS, Bockheim JG (2001) Soil development in low-arctic tundra of the northern Brooks Range, Alaska, U.S.A. Arctic Antarct Alp Res 33:78–87

Pidwirny M (2006) Periglacial processes and landforms. Fundamental of physical geography, 2nd edn, February 5, 2014. http://www.physicalgeography.net/fundamentals/10ag.html

Ping CL, Michaelson GJ, Kimble JM, Romanovsky VE, Shur YL, Swanson DK, Walker DA (2008) Cryogenesis and soil formation along a bioclimate gradient in Arctic North America. J Geophys Res 113:G03S12. doi:10.1029/2008JG000744

Seppälä M (2006) Palsa mires in Finland. In: The Finnish environment, pp 155–162. http://lustiag.pp.fi/data/pdf/2006_Seppala_PalsaMiresFinland.pdf [accessed 09-18-2014]

Soil Survey Staff (2010) Keys to soil taxonomy, 11th edn. U.S. Dept. of Agriculture, Natural Resources Conservation Service, Washington, DC

Tarnocai C, Zoltai SC (1978) Earth hummocks of the Canadian arctic and subarctic. Arctic Alpine Res 10:581–594

Tatur A (1989) Ornithogenic soils of the maritime Antarctic. Polish Polar Res 10:481–532

Walker DA, Epstein HE, Gould WA, Kelley AM, Kade AN, Knudson JA, Krantz WB, Michaelson G, Peterson RA, Ping C-L, Raynolds MK, Romanovsky VE, Shuri Y (2004) Frost-boil ecosystems: complex interactions between landforms, soils, vegetation and climate. Permafr Periglac Process 15:171–188

Walker DA, Raynolds MK, Daniels FJA, Einarsson E, Elvebakk A, Gould WA, Katenin AE, Kholod SS, Markon CJ, Melnikov ES, Moskalenko NG, Talbot SS, Yurtsev BA (2005) The circumpolar arctic vegetation map. J Veg Sci 16:267–282

Washburn AL (1956) Classification of patterned ground and review of suggested origins. Geol Soc Am Bull 67:823–866

Washburn AL (1980) Geocryology: a survey of periglacial processes and environments. Wiley, New York

Wu X, Zhao L, Chen M, Fang H, Yue G, Chen J, Pang Q, Wang Z, Ding Y (2012) Soil organic carbon and its relationship to vegetation communities and soil properties across permafrost areas of the central western Qinghai-Tibet Plateau. China Permafr Periglac Process 23:162–169

Chapter 5
Cryogenic Soil Processes

5.1 Introduction

Although soil-forming processes, such as humification, paludification, podzolization, and gleization operate in cryosols, the dominant processes are of cryogenic origin. A key misinterpretation regarding cryosols is that these cryogenic processes are geologic rather than pedogenic and belong only to the realm of geocryology (Tedrow 1966; Sokolov et al. 1980). Bockheim et al. (2006) argued that cryogenic processes involve inputs, outputs, transfers, and transformations of energy, water, and soil material and, therefore, according to classical definitions of soil-forming processes (Simonson 1959), are pedogenic (i.e., cryopedogenic).

Another misinterpretation regarding cryosols is that cryogenic processes destroy soil horizons and are "inflicted" upon "natural" soil-forming processes (Douglas and Tedrow 1960; Gerasimov 1973; Sokolov et al. 1997). In fact, Sokolov and others (1997) claimed that "cryogenic processes … do not result in soil formation" (p. 7) and cryosols are characterized by the "absence of well developed pedogenic horizons and features" (p. 4). Tedrow (1968) identified two sets of processes acting contemporaneously on polar soils: a *pedologic* process that gives rise to a "genetic" morphology and a *geologic* process that tends to disrupt any acquired morphology (Douglas and Tedrow 1960; Tedrow 1968). Tedrow (1968) referred to these destructive elements as "cannibalization" and viewed the so-called "natural" soil-forming processes as resulting in soil horizons more or less parallel to the ground surface, and geologic processes as resulting in irregular and broken horizons reflective of a "negative" process contrary to soil formation.

Bockheim et al. (2006) posited that cryopedogenic processes are "natural" and characteristic of permafrost-affected soils. An analogous situation is the "vertization" process leading to the development of vertisols. The irregular and broken horizons that are common to cryosols are the natural product of cryopedogenic processes, such as cryoturbation, freeze–thaw, frost heaving, cryogenic sorting, thermal cracking, and ice segregation (Bockheim and Tarnocai 1998). These processes are characteristic

J.G. Bockheim, *Cryopedology*, Progress in Soil Science,
DOI 10.1007/978-3-319-08485-5_5

of permafrost regions and result in soils that have markedly different properties than those not influenced by cryopedogenic processes (Makeev and Kerzhentsev 1974; Hendershot 1985; Dobrovol'skiy 1996). For example, podzolic soils underlain by permafrost are "genetically … the product of the podzolic process in combination with the cryogenic process" (Kuzmin and Sazonov 1965, p. 1272). Permafrost has actually induced podzolization in the Transbaikal region through its control on hydrologic and thermal regimes. In other environments, like those of central and southwestern Yakutia, permafrost, along with low precipitation, suppresses podzolization by limiting leaching of weathering products (Zolnikov et al. 1962).

Many aspects of the cryopedogenic process are "positive" in the traditional sense, including size reduction of particles, arrangement of soil particles, formation of soil aggregates, disintegration of rocks, and ice–salt exclusion (Tedrow 1968; Makeev 1981; Marion 1995). Gerasimov (1975) identified deformation, which included cryogenic processes, as key elementary soil processes. Cryopedogenic processes can be observed at the landscape scale by the presence of patterned ground and at the pedon scale by cryoturbation (transfer), cryodesiccation (transfer), and ice segregation (transfer/transformation). At the microscopic level, these processes are manifested by characteristic platy, blocky, or vesicular macrostructures and banded and orbiculic microstructures (Gerasimova et al. 1992; Fox 1994; Rusanova 1996, 1998).

5.2 Cryopedologic Processes

Cryogenic soil processes may be viewed from a landscape scale, a pedon scale, and a microscopic scale. At the landscape scale, cryopedogenesis is evidenced by a variety of patterned ground features (Figs. 4.5 and 4.6). Figure 5.1 depicts the evolution of cryogenic processes during soil development in a permafrost region, i.e., at the pedon scale. Cold temperatures are the forcing factor. In the early stages of soil development, yield an immature soil derived from relatively unaltered parent materials underlain by permafrost. As the soil undergoes development, the presence of permafrost and the resulting active layer enable processes such as frost heaving, cryoturbation, thermal cracking, and ice segregation to occur. These processes yield irregular and broken soil horizon boundaries, the development of platy and massive structures. The soil horizons that form reflect not only these cryopedologic processes but also the imprint of other pedologic processes such as humification (development of organic layers and organic-enriched A horizons), cambisolization (development of color or structural B horizons), gleization (restriction of water flow by the underlying permafrost), podzolization, etc.

At the microscopic scale, thin sections can show compaction from cryodesiccation, displacement from cryoturbation and related processes, and pore-formation from ice segregation (Fig. 5.2). These features are described using terminology from the sub-discipline of soil micromorphology (Bullock et al. 1985).

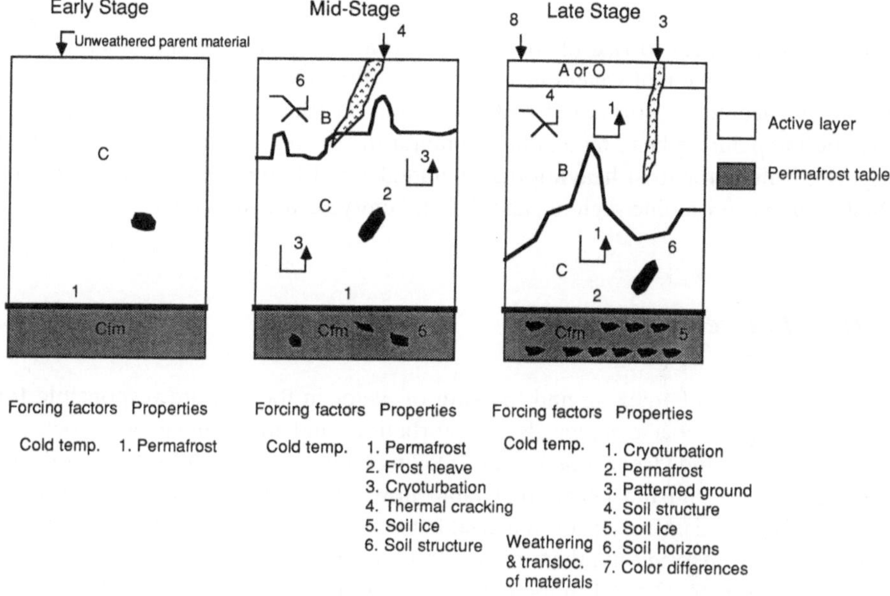

Fig. 5.1 Morphogenetic development of a permafrost-affected soil

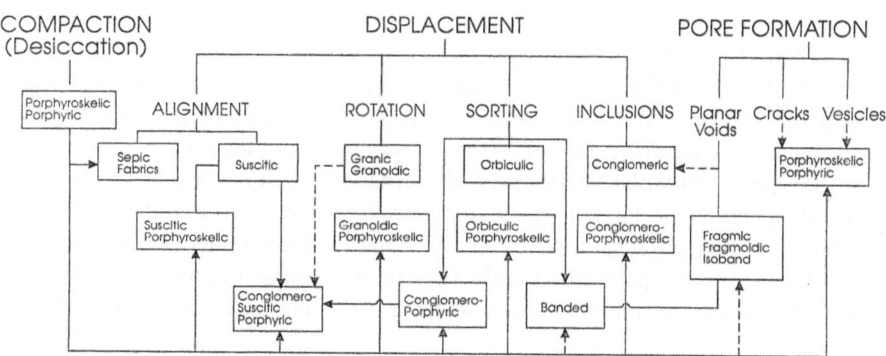

Fig. 5.2 Cryogenic processes evidenced in thin sections (Fox 1985)

Cryogenic processes include frost heaving, cryoturbation, dilatancy, cryodesiccation, and ice segregation, each of which will be discussed in the following sections. Detailed reviews of these processes have been provided by Tedrow (1977), Washburn (1980), Yershov and Williams (2004), French (2007); van Vliet-Lanoë 2010).

5.2.1 *Frost Heaving*

Frost heaving refers to rise of the ground surface or of coarse fragments due to volumetric expansion of water during the formation of segregation ice. Key mechanisms are needle-ice formation and cryoturbation (Sect. 5.2.3). Large pressures may build up during heaving that have a major impact on structures. French (2007) reported values for frost heaving that included 1.5–14 cm/year for arctic regions, 2–30 cm/year for alpine regions, and 0.4–4.0 cm/year for Antarctica.

5.2.2 *Freeze-Thaw Cycles*

Repeated cycles of freezing and thawing of water in the soil are responsible for frost heaving of coarse materials, cryoturbation, and mechanical or cryoclastic weathering. During "freeze-back", the freezing fronts move both from the soil surface downward and the permafrost table upward. As this happens, moisture is removed from the unfrozen soil material between the two fronts (cryodesiccation). French (2007) reported numbers of freeze-thaw cycles (surface or 1–2 cm depth) that included from 23 to 94 in the Arctic, 9 to 50 in alpine regions, and 19 to 42 in Antarctica.

5.2.3 *Cryoturbation*

Cryoturbation is often recognized as the most important cryopedologic process in cryosols (Tedrow 1962; van Vliet-Lanoë 1988; Bockheim and Tarnocai 1998; Bockheim et al. 1998a; Bockheim 2007; Jelinski 2013). Cryoturbation is a collective term used to describe all soil movements due to frost action. Cryoturbation is evidenced by (1) irregular horizons, (2) deformation of textural bands from deposition in the parent material, (3) broken horizons, (4) involutions, (5 the accumulation of fibrous or partially decomposed organic matter concentrated in the transition zone or on top of the permafrost table, (6) oriented coarse fragments, (7) silt caps from vertical sorting, and (8) upwarping of sediments adjacent to sand- or ice-wedges (Fig. 3.1).

Cryoturbation is a complicated process that involves winter freezing of the active layer that enables ice lenses to form accompanied by loss of water from the adjoining soil and summer thawing resulting in subsidence and dilation (Fig. 5.3). Although this model was developed by Mackay (1980) to explain the formation of earth hummocks in arctic Canada, it illustrates the cryoturbation process. Examples of cryoturbation in cryosols are shown in Fig. 5.4.

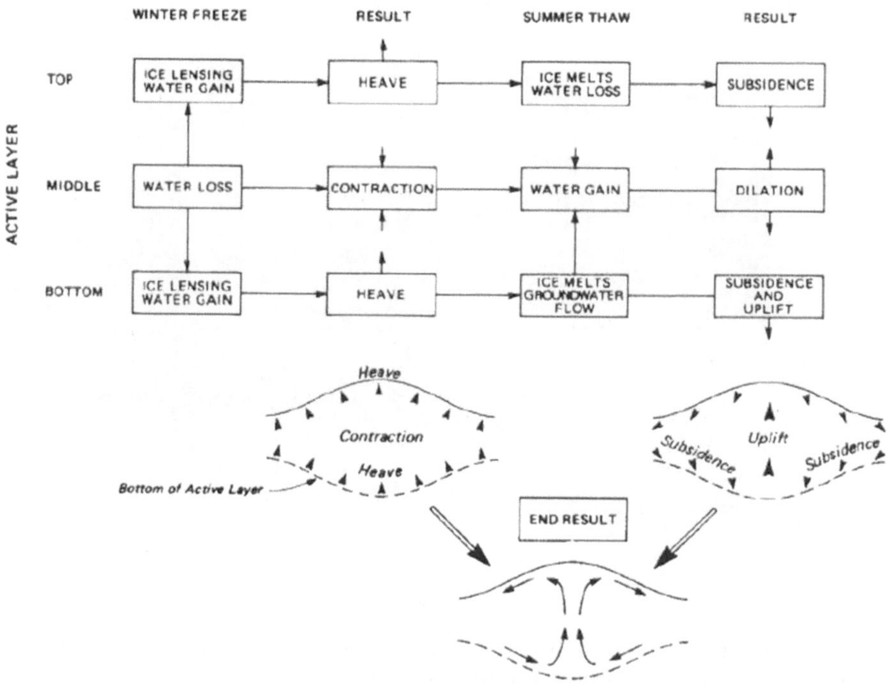

Fig. 5.3 A model illustrating cryoturbation during growth of an earth hummock in arctic Canada (Mackay 1980)

5.2.4 *Dilatancy*

In the literature there is confusion among the terms dilatancy and thixotrophy (Alexander 1992) and liquefaction. Dilatancy is defined as the property of dilating or expanding, particularly in granular materials that expand due to the increase in space between rigid particles upon displacement of the particles. In contrast, thixotrophy is defined as the property of certain gels becoming fluid when agitated and reverting back to a gel when left to stand. Soil liquefaction is the phenomenon whereby a saturated or partially saturated soil substantially loses strength and stiffness in response to an applied stress, often during an earthquake. Bockheim et al. (1998b) observed that the Bg horizon of soils in moist nonacidic tundra often displayed dilatancy due to the abundance of silt and water in the active layer.

5.2.5 *Cryodesiccation*

The primary evidence for cryodesiccation is the presence of cracks that often extend from the soil surface through the active layer into the near-surface permafrost. Cryodesiccation is also evidenced by dry zones with the soil that may contribute to

Fig. 5.4 Examples of cryoturbation in arctic Alaska, including (**a**) buried and redistributed soil organic carbon in sandy alluvium; (**b**) patches of redistributed SOC in sandy alluvium; (**c**) cryoturbation in a Spodosol-like soil (Psammoturbel) in sandy alluvium; and (**d**) SOC concentrated in near-surface permafrost of a drained thaw-lake basin (Bockheim 2007)

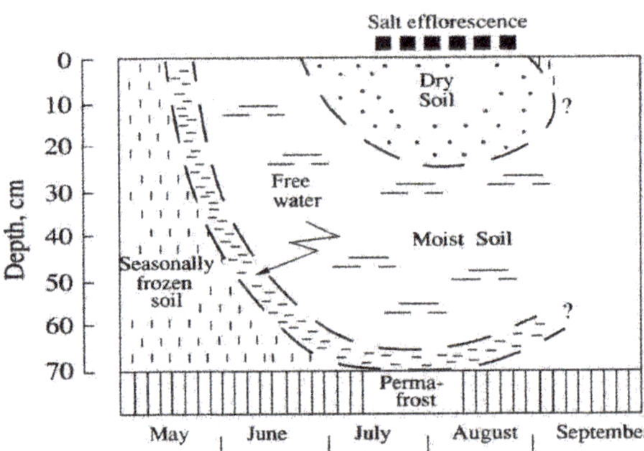

Fig. 5.5 A generalized depiction of seasonal changes in soil thaw and moisture status of a Polar Desert soil (Tedrow 2004)

the development of salt efflorescence on the soil surface. This process is evident is soils of the arctic and the Antarctic. In arctic soils, cryodesiccation occurs as water is drawn from unfrozen areas during freeze back (Fig. 5.5), and is responsible for the platy or blocky structures or massive, structureless conditions of cryosols

Fig. 5.6 Platy (*top*) and blocky (*bottom*) structures resulting from cryodesiccation

(Fig. 5.6). In ultraxerous cryosols (Anhyorthels and Anhyturbels) of Antarctica, cryodesiccation is responsible for cracks that develop following melting of snow as well as thermal contraction of soils with sand-wedge polygons (Fig. 5.7).

5.2.6 Ice Segregation

Ice segregation occurs when liquid moisture is transferred to other regions in the soil during freezing. When ice forms in sandy soils, practically no migration of moisture occurs, and interstitial ice generally develops during freezing. In contrast, the freezing of finer textured soils induces the formation of inclusions of pure segregation ice as lenses, layers, reticulations, etc. (see Figs. 2.8 and 3.2).

5.2.7 Gelifluction

Gelifluction refers to the slow downslope flow of unfrozen earth materials on a frozen substrate. Therefore, it is a mass-wasting process and not a cryopedogenic process, but it is strongly influenced by frost. The term "solifluction" normally is restricted to downslope flow of unfrozen earth materials due to seasonal frost rather than to permafrost. Gelifluction is strongly influenced by the amount and distribution of segregation ice (Matusoka, 2001). Rapid movement is often restricted to a superficial layer upper 5–10 cm thick, especially in fine-textured soils. Gelifluction can occur on slopes of less than 10°, and common surface velocities range from 1 to 3 cm/year.

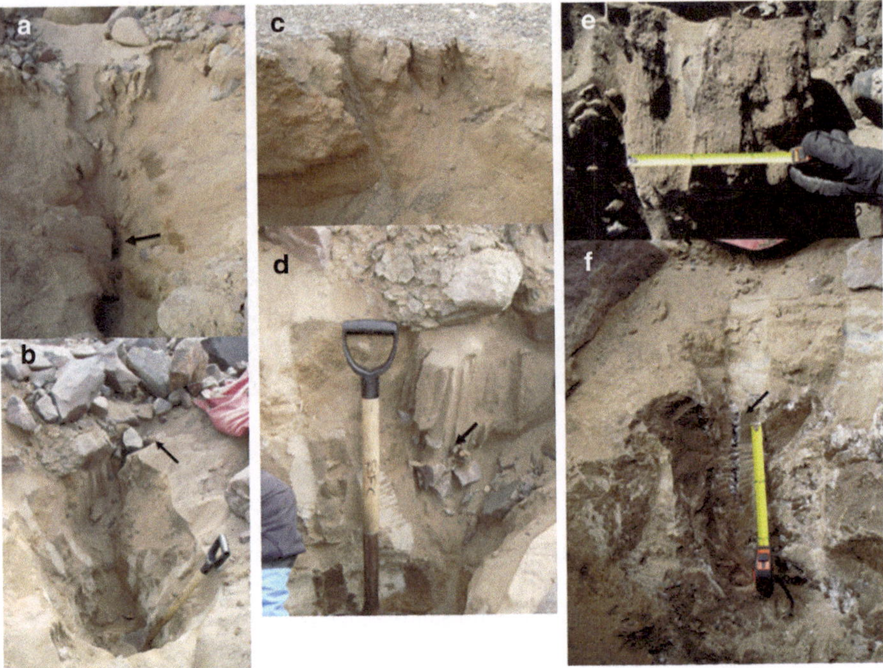

Fig. 5.7 Cryodesiccation in Beacon Valley, Antarctica as reflected by (**a**) a sand wedge with a contraction fissure (*arrow*); (**b**) an exposed fissure (*arrow*) surrounded by ice-bonded permafrost; (**c**) a series of inactive fissures (*arrows*) from cryodesiccation; (**d**) vertical laminations of sand as columns on stones (*arrow*) in a sand wedge; (**e**) a lateral view of rivulets of flowing sand along a contraction cracks; and (**f**) ice veins (*arrow*) in fissures of a sand wedge (Bockheim et al. 2009)

5.3 Cryopedologic Processes and Soil Classification

All of the soil classification schemes developed for polar regions utilize cryopedogenic processes in some fashion. Tedrow (1973) viewed soil development and patterned ground formation as "separate entities." However, he recognized certain relationships between the two. Tedrow's (1973) zonal scheme had four levels of classification that did not included patterned ground. For future detailed soil mapping, Tedrow left open the idea of using both the genetic soil and a variety of patterned ground.

As will be seen in Chap. 6, the soil classification systems developed in the USA and Canada and by the WRB all divide cryosols/gelisols using cryoturbation at the highest level. In Russia soil with strongly cryoturbated horizons are classified as cryozems (Sokolov et al. 1997).

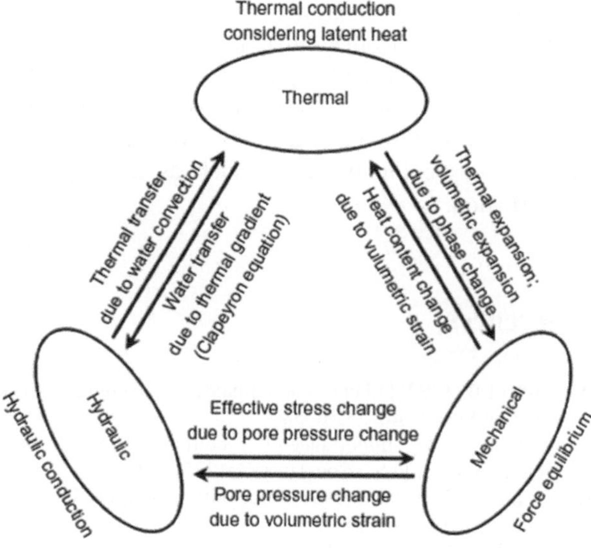

Fig. 5.8 Thermo-hydro-mechanical interaction mechanism in freezing soils (Thomas et al. 2009)

5.4 Modeling Cryogenic Soil Processes

Thomas et al. (2009) used a numerical approach to model cryogenic processes in permafrost-affected soils (Fig. 5.8). The model considered soil freezing and ice segregation. During two-sided freezing, ice segregation took place in a closed system, with ice lenses accumulating at the base of the active layer and near the ground surface, leaving an intervening ice-poor zone. Numerical modeling was able to represent the development of both the thermal field and ice segregation observed in the physical models. The significance of this contrasting ground ice distribution is considered in the context of thaw-related slow mass movement processes (gelifluction).

5.5 Summary

Although soil-forming processes, such as humification, paludification, podzolization, and gleization operate in cryosols, the dominant processes are of cryogenic origin. Cryogenic processes involve inputs, outputs, transfers, and transformations of energy, water, and soil material and, therefore, according to classical definitions of soil-forming processes. Cryogenic processes include frost heaving, cryoturbation, dilatancy, cryodesiccation, and ice segregation. These processes are evidenced at the landscape scale in the form of patterned ground, at a pedon scale in the form of cryoturbation, dilatancy, ped forms, and cryodesiccation features, and at the microscopic scale from compaction, displacement of plasma or skeletal grains, and pore formation. Cryopedologic processes are utilized in the definition of cryosols or gelisols and at the second highest level in distinguishing whether or not cryoturbation is observed in mineral soils. Numerical models of cryopedogenesis have focused on specific processes such as heaving, formation of segregation, and cryoturbation.

References

Alexander EB (1992) Dilatancy and thixotrophy: commonly misused terms and confused phenomena. Soil Surv Horiz 33:36–38

Bockheim JG (2007) Importance of cryoturbation in redistributing organic carbon in permafrost-affected soils. Soil Sci Soc Am J 71:1335–1342

Bockheim JG, Tarnocai C (1998) Recognition of cryoturbation for classifying permafrost-affected soils. Geoderma 81:281–293

Bockheim JG, Walker DA, Everett LR (1998a) Soil carbon distribution in nonacidic and acidic tundra of arctic Alaska. In: Lal R, Kimble JM, Follett RF, Stewart BA (eds) Soil processes and the carbon cycle. CRC Press, Boca Raton, pp 143–155

Bockheim JG, Walker DA, Everett LR, Nelson FE, Shiklomanov NI (1998b) Soils and cryoturbation in moist nonacidic and acidic tundra in the Kuparuk River basin, arctic Alaska, U.S.A. Arctic Alp Res 30:166–174

Bockheim JG, Mazhitova G, Kimble JM, Tarnocai C (2006) Controversies on the genesis and classification of permafrost-affected soils. Geoderma 137:33–39

Bockheim JG, Kurz MD, Soule SA, Burke A (2009) Genesis of active sand-filled polygons in lower and central Beacon Valley, Antarctica. Permafr Periglac Process 20:295–308

Bullock P, Fedoroff N, Jongerius A, Stoops G, Tursina TE, Babel U (1985) Handbook for thin section description. WAINE Research Publications, Albrighton, Wolverhampton

Dobrovol'skiy VV (1996) Basic features of the geochemistry of arctic soil formation. Eurasian Soil Sci 28:218–230

Douglas LA, Tedrow JCF (1960) Tundra soils of arctic Alaska. 7th International Congress of Soil Science, Madison, WI, vol 41, pp 291–299

Fox CA (1985) Micromorphology of an Orthic Turbic cryosol – a permafrost soil. In: Bullock P, Murphy CP (eds) Soil micromorphology, vol 2, Soil genesis. A B Academic Publishers, Berkhamsted, pp 699–705

Fox CA (1994) Micromorphology of permafrost-affected soils. In: Kimble JM, Ahrens RJ (eds) Proceedings of the meeting on the classification, correlation, and management of permafrost-affected soils. USDA Soil Conservation Service, National Soil Survey Center, Lincoln, pp 51–62

French HM (2007) The periglacial environment, 3rd edn. Wiley, New York

Gerasimov IP (1973) Elementary soil-forming processes as a basis for the genetic diagnostics of soils. Pochvovedenie 5:102–113 (In Russian)

Gerasimov IP (1975) Experiment in genetic diagnosis of the soils of the USSR on the basis of elementary soil processes. Sov Soil Sci 7:257–263

Gerasimova MI, Gubin SV, Shoba SA (1992) Micromorphological features of the USSR Zonal soils. Publishing Service of the Pushchino Scientific Center, RAS. 215 pp (In Russian)

Hendershot WH (1985) Comparison of Canadian and American classification systems for some arctic soils of the Ungava–Labrador Peninsula. Can J Soil Sci 65:283–291

Jelinski NA (2013) Cryoturbation in the central Brooks Range, Alaska. Soil Horiz. doi:10.2136/sh13-01-0006

Kuzmin VA, Sazonov AG (1965) Podzolic soils of the Chara River basin (northern Transbaikal region). Sov Soil Sci 11:1268–1276

Mackay JR (1980) The origin of hummocks, western Arctic coast, Canada. Can J Earth Sci 17:996–1006

Makeev OV (1981) Facies of soil cryogenesis and associated soil profile features. Nauka, Moscow, 87 pp (In Russian)

Makeev OV, Kerzhentsev AS (1974) Cryogenic processes in the soils of northern Asia. Geoderma 12:101–109

Marion GM (1995) Freeze–thaw processes and soil chemistry, vol 95–12, Special report. US Army Corps of Engineers, Cold Regions Research and Engineering Laboratory, Hanover, 23 pp

Matsuoka N (2001) Solifluction rates, processes and landforms: a global view. Earth-Sci Rev 55:107–134

Rusanova GV (1996) Micromorphology of East-European subarctic soils. Eurasian Soil Sci 29:715–724

Rusanova GV (1998) Micromorphology of human-Affected soils. Ural Division of Russian Academy of Sciences Publishing, Ekaterinburg, 160 pp (In Russian)

Simonson RL (1959) Outline of a generalized theory of soil genesis. Soil Sci Soc Am Proc 23:152–156

Sokolov IA, Chigir VG, Alifanov VM, Khudyakov OI, Gugalinskaya LA, Fominykh LA, Gilichinsky DA, Maksimovich SV (1980) Concepts, terminology, and classification problems in the study of freezing soils. Sov Soil Sci 12:666–673

Sokolov IA, Naumov YM, Konyushkov DY (1997) The placement of cryosols in the world reference base for soil resources. In: Cryosols in classification hierarchy. Russian Academic Science/Nauka, Novosibirsk, pp 4–10

Tedrow JCF (1962) Morphological evidence of frost action in Arctic soils. Biul Peryglacjalny 11:343–352

Tedrow JCF (1966) Polar desert soils. Soil Sci Soc Am Proc 30:381–387

Tedrow JCF (1968) Pedogenic gradients of the polar regions. J Soil Sci 19:197–204

Tedrow JCF (1973) Polar soil classification and the periglacial problem. Biul Peryglacjalny 22:285–294

Tedrow JCF (1977) Soils of the polar landscapes. Rutgers University Press, New Brunswick

Tedrow JCF (2004) Polar Desert soils in perspective. Eurasian Soil Sci 37:443–450

Thomas HR, Cleall P, Li Y-C, Harris C, Kern-Luetschg M (2009) Modelling of cryogenic processes in permafrost and seasonally frozen soils. Géotechnique 59:173–184

Van Vliet-Lanoë B (1988) The significance of cryoturbation phenomena in environmental reconstruction. J Quat Sci 3:85–96

Van Vliet-Lanoë B (2010) Frost action. In: Stoops G, Marcelino V, Mees F (eds) Interpretation of micromorphological features of soils and regoliths. Elsevier, New York, pp 81–108

Washburn AL (1980) Geocryology: a survey of periglacial processes and environments, 2nd edn. Wiley, New York, 406 pp

Yershov ED, Williams PJ (2004) General geocryology. Cambridge University Press, Cambridge

Zolnikov VG, Elovskaya LG, Teterina LV, Chrnyak EI (1962) Soils of the Vilui River Basin and their use. USSR Academy of Sciences Publishing, Moscow, 204 pp (In Russian)

Chapter 6
Classification of Cryosols

6.1 Zonal Soil Classification Systems

From 1900 to 1960, soil classification systems in Russia and the USA were genetic, i.e., based on presumed soil-forming processes, rather than technical or natural, i.e., based on soil properties. For example, in the scheme by Baldwin et al. (1938), soils were subdivided into zonal soils, well-developed soils that were in equilibrium with the climate and vegetation, intrazonal soils, well-developed soils that were influenced by some local factor such as drainage or parent material, and azonal soils, those which lacked development because they were young or lacked "pedogenic inertia" (Bryan and Teakle 1949). Baldwin et al. (1938) recognized zonal soils in the cold zone that included Tundra soils and Subarctic Brown Forest soils and intrazonal soils that included Alpine meadow soils.

Zonal systems were also employed in the polar regions by Tedrow (1968, 1977, 1991), in the arctic by Ugolini (1986), and in the Antarctic by Campbell and Claridge (1969) and Bockheim and Ugolini (1990). Tedrow (1968) prepared a schematic arrangement of major genetic soils in the polar regions (Table 6.1). The soils were arrayed along a hypothetical moisture gradient from the arctic tundra to the High Arctic and then to the Cold Desert of Antarctica. The Arctic Brown was the zonal soil in arctic tundra in the region of continuous permafrost; the Polar Desert soil dominated poorly vegetated High Arctic landscapes underlain by continuous permafrost; and the Cold Desert soil was the major soil in ice-free areas of Antarctica. In Antarctica Campbell and Claridge (1969) described the zonal soils of Antarctica as frigic soils (Table 6.2). Frigic soils were divided according to the available moisture status, which strongly influenced soil development: ultraxerous/weakly developed, xerous/moderately developed, and subxerous/strongly developed. Later studies showed that the most strongly developed soils were Miocene-aged soils along the edge of the polar plateau, and subxerous soils along the coast were poorly developed because they were of Late Glacial Maximum age and younger and were strongly cryoturbated.

J.G. Bockheim, *Cryopedology*, Progress in Soil Science,
DOI 10.1007/978-3-319-08485-5_6

Table 6.1 Schematic arrangement of major genetic soils in the polar regions

Soil zone	Deep, well-drained soils	Gley soils	Organic soils
Arctic Brown	Arctic Brown	Tundra	Bog
	Minimal Podzol in southern sectors		
Polar Desert	Polar Desert	Tundra	Bog
	Isolated Arctic Brown soil in southern sectors		
Cold Desert	Cold Desert		
	Evaporite (saline) soils of the Cold Desert		

Tedrow (1968)

Table 6.2 Zonal classification of frigic soils of Antarctica

		Parent material	
Available moisture Status	Degree of development	Material	Composition
Zonal soils (Frigic soils)			
Ultraxcroos	Weakly developed	Alluvium	Greywacke
Xerous	Moderately developed	Colluvium	Schist
Subxerous	Strongly developed	Moraine	Granite
		Massive rock, etc.	Dolerite, etc.
Intrazonal soils			
Soils formed under the influence of saline groundwater	Evaporitc Soils		
	Algal peats		
Soils with a dominant organic constituent	Avian soils (ornithogenic soils)		
Soils formed under the influence of hydrothermal waters	(Hydrothermal soils)		
Azonal soils			
Soils of recent beaches, fans, streambeds, etc.	Recent soils		

Source: Campbell and Claridge (1969)

Table 6.3 shows Tedrow's (1991) linkage of soils in the polar regions. The Cold Desert (Antarctica) and Polar Desert (Arctic) Soil Zones were subdivided into second order (zonality and soil moisture availability), third order (texture and mineral properties of the parent material), fourth order (wetness factor; not applicable to Antarctica), and fifth order (related to patterned ground). Although the zonal systems are useful in relating soils of the Arctic and Antarctic, they have been used only to a limited extent in mapping soils of the polar regions.

6.2 Natural Soil Classification Schemes

The first technical or natural soil classification system was established in Canada in 1978. cryosols were identified as one of what was to become 10 soil orders. cryosols were recognized as forming in either mineral or organic materials that have

Table 6.3 Zonal classification of soils in Antarctica (top) and the High Arctic (bottom)

First order	Second order	Third order	Fourth order	Fifth order
Cold Desert Soil Zone	Ahumic (Frigic) soils	Separations based on textural and mineral properties of the parent material, etc.	Not applicable for the mainland of Antarctica	Soil type + patterned ground
	Ultraxerous			
	Xerous			
	Subxerous			
	Ahumisol			
	Evaporite soils			
	Omilhogenic (Avian)			
	Soils			
	Other soils			
	Protoranker			
	Algae peats			
	Hydrothermal soils			
	Regosols (recent soils)			
	Lithosols			
Polar Desert Soil Zone	Well-drained soils	Separations based on textural and mineral properties of the parent material, etc.	Soil type + wetness factor (applies mainly to Tundra and Bog soils)	Soil type + wetness factor + patterned ground
	Polar Desert soil			
	Arctic Brown soil			
	Mineral gley soils			
	Upland Tundra			
	Meadow Tundra			
	Soils of the hummocky ground			
	Soils of the Polar Desert–Tundra interjacence			
	Organic soils			
	Bog soils			
	Other soils			
	Regosols			
	Lithosols			
	Soils of the solifluction slopes (may be a form of gley soil but usually well drained)			
	Rendzina			

Tedrow (1991)

permafrost either within 1 m of the surface or within 2 m if the pedon has been strongly cryoturbated laterally with the active layer. The Crysolic order was divided into three great groups: Turbic cryosols, Static cryosols, and Organic cryosols, based on the degree of cryoturbation and the nature of soil material (Table 6.4). This system has undergone two subsequent revisions, and a fourth revision is in progress.

Table 6.4 The cryosolic order in the Canadian soil classification system

Cryosolic order			
	Turbic cryosol	Static cryosol	Organic cryosol
Soil	Mineral	Mineral	Organic
Cryoturbation	Marked, usually patterned ground	None	None
Permafrost	Within 2 m of surface	Within 1 m of surface	Within 1 m of surface

Soil Classification Working Group (1998)

Table 6.5 Distribution of cold soil taxa by soil temperature class

Soil temperature class	MAST, 50 cm (°C)	Permafrost zonation	Dominant soil taxa
Hypergelic	−10 or colder	Continuous	Gelisols
Gelic	−4 to −10	Discontinuous	Gelisols
Subgelic	+1 to −4	Sporadic	Gelaquents, Gelifluvents, Golorthents,
			Gelepts, Cryo(Geli-) fibrists, Gellols,
			Gelods, Gelands
Frigid	<8	Seasonal frost	Cryids, Cryalfs, Cryands, Cryolls,
			Cryods, Cryepts, Cryerts, Cryaquents,
			Cryofluvents, Cryorthents, Cryopsamments,
			Cryofibrists, Cryofolists, Cryohemists,
			Cryosaprists

The first technical soil classification system in the USA went through a series of six approximations before yielding the *Seventh Approximation* (Soil Survey Staff 1960). This scheme recognized permafrost and classified soils previously identified as Tundra soils and Subarctic Brown Forest soils as Entisols (Cryaquents and Cryudents), Inceptisols (Cryaquepts, Cryandepts, Cryumbrepts, and Cryochrepts), and Histosols (order in development stage). In 1975 *Soil Taxonomy* (Soil Survey Staff 1975) recognized permafrost-affected soils in Pergelic subgroups of Entisols (Cryorthents and Cryopsamments), Histosols (Cryofibrists, Cryohemists, and Cryosaprists), Inceptisols (Cryaquepts, Cryochrepts, and Cryumbrepts), and Spodosols (Cryaquods, Cryohumods, and Cryorthods).

The second edition of *Soil Taxonomy* (Soil Survey Staff 1999) included Gelisols, the permafrost-affected soils. These were divided into three suborders: Histels, organic soils with permafrost within 1 m of the surface; Turbels, mineral soils with cryoturbation and permafrost within 2 m of the surface; and Orthels, mineral soils with no cryoturbation and permafrost within 1 m of the surface. Great groups were based on diagnostic horizons. The distribution of cold soil taxa by soil temperature class is shown in Table 6.5. Gelisols are ubiquitous in areas where the soil-temperature class (STC) is hypergelic (−10 °C or colder) and are predominant in areas where the STC is gelic (−4 to −10 °C). However, in areas with a subgelic STC (+1 to −4 °C), permafrost is below the 1 or 2 m required depth for Gelisols, and the soils are classified in Gel- suborders or great groups. A frigid STC (1–8 °C) includes soils with seasonal frost; these soils lack permafrost are classified in Cry- suborders and great groups.

There are 52 Gelisols soil series in the NRCS SSURGO database for the USA, including all three suborders, 9 great groups, and 14 subgroups (Table 6.6). It is of interest that only 14 (13 %) of the 107 possible subgroups allowed in the Gelisols are represented by identified soil series; and 38 (73 %) of the 52 soil series occur in three subgroups: Typic Historthels, Typic Aquiturbels, and Typic Histoturbels. However, the database contains additional Gelisols from Russia and Antarctica.

In 2006 a cryosol subgroup was recognized as one of 32 such subgroups in the WRB (IUSS WRB Working Group 2006). cryosols were defined as mineral soils with permafrost within 100 cm of the ground surface and were considered to be analogous to Gelisols in ST, cryozems in the Russian scheme, and Polar Desert soils in the zonal system of Tedrow (1977). Detailed information about cryosols in the WRB is given in Tarnocai et al. (2004).

Table 6.6 Classification of soils within the Gelisol order and current soil series recognized in the USA

Suborder	Great group	Subgroup	Soil series recognized
Histels	Folistels	Lithic	
		Glacic	Fels
		Typic	Peluk
	Glacistels	Hemic	
		Sapric	
		Typic	
	Fibristels	Lithic	
		Terric	
		Fluvaquentic	
		Sphagnic	
		Typic	Lemata
	Hemistels	Lithic	
		Terric	Wrangell
		Fluvaquentic	Haggard
		Typic	Bolio, Bonot
	Sapristels	Lithic	
		Terric	
		Fluvaquentic	
		Typic	

(continued)

Table 6.6 (continued)

Suborder	Great group	Subgroup	Soil series recognized
Orthels	Aquorthels	Lithic	
		Glacic	
		Sulfuric	
		Ruptic-Histic	
		Andic	
		Vitrandic	
		Salic	
		Psammentic	
		Fluvaquentic	Happy
		Typic	Dotlake, Kuslinad, Mendna
	Argiorthels	Lithic	
		Glacic	
		Natric	
		Typic	
	Haplorthels	Lithic	
		Glacic	
		Fluvaquentic	
		Folistic	
		Aquic	
		Fluventic	
		Typic	Hogan
	Historthels	Lithic	
		Glacic	
		Fluvaquentic	
		Fluventic	
		Ruptic	
		Typic	Boldrin, Copper River, Dadina, Dirrant, Goodpaster, Klanelneechena, Klawasi, Kuslina, Kuswash, Mendottna, Minchumina, Owhat, Tolson
	Mollorthels	Lithic	
		Glacic	
		Vertic	
		Andic	
		Vitrandic	
		Folistic	
		Cumulic	
		Aquic	
		Typic	
	Psammorthels	Lithic	
		Glacic	
		Spodic	
		Typic	
	Umbrorthels	Lithic	
		Glacic	
		Vertic	
		Andic	

(continued)

Table 6.6 (continued)

Suborder	Great group	Subgroup	Soil series recognized
Turbels		Vitrandic	
		Folistic	
		Cumulic	
		Aquic	
		Typic	
	Aquiturbels	Lithic	
		Glacic	
		Sulfuric	
		Ruptic-Histic	Dustin, Frostcircle
		Psammentic	
		Typic	Bradway, Browne, Chatanika, Chet Laske, Kindanina, Klasi, Mankomen, Ninchuun, Tanana, Tatlanika, Tetlin
	Haploturbels	Lithic	
		Glacic	
		Folistic	
		Aquic	
		Typic	Babel
	Hisoturbels	Lithic	
		Glacic	
		Ruptic	
		Typic	Mosquito, Swillna, Ester, Goldstream, Iksgiza, Saulich, Steps, Strelna, Tanacross, Teggiuq, Totatlanika, Turbellina, Windy Creek
	Molliturbels	Lithic	
		Glacic	
		Vertic	
		Andic	
		Vitrandic	
		Folistic	
		Cumulic	
		Aquic	
		Typic	
	Psammoturbels	Lithic	
		Glacic	
		Spodic	
		Typic	
	Umbriturbels	Lithic	
		Glacic	
		Vertic	
		Andic	
		Vitrandic	
		Folistic	
		Cumulic	
		Aquic	
		Typic	

In Russia there are a large number of soil classification systems that differ significantly among each other. Not all soils containing permafrost are classified as the equivalent of cryosols in Russia (Mazhitova 2004). Soils that are strongly cryoturbated are classified as cryozems; these soils normally have a shallow active layer.

Table 6.7 compares soil taxa in ST, Canada, the WRB, and the Russian systems as interpretation by Mazhitova (2004). Gelisol great groups in ST are comparable to

Table 6.7 A comparison of cryosol soil taxa among global and country systems

Soil taxonomy (Soil Survey Staff 2010)	Canada (Soil Classification Working Group 1998)	WRB (IUSS Working Group, WRB 2006)	Russian (Mazhitova 2004)
Great group	Subgroup	Soil group	
Histels	Organic Cryosols	Histic Cryosols	
Folistels	–	Folic Cryosols	Dry-Peaty
Glacistels	Glacic Organic Cryosols	Glacic Cryosols	–
Fibristels	Fibric Organic Cryosols	Fibric Cryosols	Oligotrophic Peat
Hemistels	Mesic Organic Cryosols	Hemic Cryosols	Oligotrophic Peat
Sapristels	Humic Organic Cryosols	Sapric Cryosols	Eutrophic Peat
Orthels	Static Cryosols	[Non-Turbic Cryosols]	
Historthels	Histic Static Cryosols	Histic Cryosols	Peat Gleyzems, Peat Cryozems
Aquorthels	Gleysolic Static Cryosols	Gleyic Cryosols	Gleyzems
Anhyorthels	–	–	–
Mollorthels	Brunisolic Eutric Static Cryosols	Mollic Cryosols	–
Umbrorthels	Brunisolic Dystric Static Cryosols	Umbric Cryosols	–
Argiorthels	Luvisol Static Cryosols	Luvic Cryosols	–
Psammorthels	Regosolic Static Cryosols	Arenic Cryosols	Podzols
Haplorthels	Orthic Static Cryosols	Haplic Cryosols	Cryozems, Podburs
Turbels	Turbic Cryosols	Turbic Cryosols	
Histoturbels	Histic Turbic Cryosols	Histic-turbic Cryosols	Peat Gleyzems, Peat Cryozems
Aquiturbels	Gleysolic Turbic Cryosols	Gleyic-turbic Cryosols	Gleyzems
Anhyturbels	–	–	–
Molliturbels	Brunisolic Eutric Turbic Cryosols	Molli-turbic Cryosols	–
Umbriturbels	Brunisolic Dystric Turbic Cryosols	Umbri-turbic Cryosols	–
Psammoturbels	Regosolic Turbic Cryosols	Areno-turbic Cryosols	Podzols
Haploturbels	Orthic Turbic Cryosols	Haplo-turbic Cryosols	Cryozems, Podburs

Table 6.8 Classification schemes for permafrost-affected soils

Person or country	Year	Approach	Main features
Tedrow, J.C.F.	1977 (1968, 1991)	Genetic	Geographic – pedogenetic gradients
Campbell, I.B. & Claridge, G.G.C.	1987 (1969)	Genetic	Soil climate, weathering stage
Canadian Soil Classification System	1998	Natural	Cryosolic (permafrost); organic vs. mineral; cryoturbation; soil properties
Russian Federation (Fridland; Shiskov, etc.)	1982	Genetic	8 soil orders may have permafrost; Cryozems all contain permafrost, lack diagnostic horizons
Soil Taxonomy (USA)	2010 (1999)	Natural	Gelic materials w/ permafrost; organic vs. mineral; cryoturbation; diagnostic soil horizons
WRB for Soil Resources	2006	Natural	Cryic horizon (permafrost); diagnostic soil horizons
Chinese Soil Taxonomic System (Zhang & Gong)	2001	Natural	Subdivision of Primosols, Gleyosols, Cambosols; · permafrost

cryosol subgroups in the Canadian system, and cryosol soil groups in the WRB. Table 6.8 lists the cryosol classification systems described herein.

6.3 Diagnostic Horizons in Cryosols

In *Soil Taxonomy* (ST) soils are classified according to diagnostic horizons (epipedons and subsurface horizons) and characteristics or properties that may be used for mineral material, organic material, or both.

In ST Gelisols, which are analogous to cryosols in the Canadian and WRB systems, are defined on the basis of the presence of permafrost within 1 m of the surface or the presence of gelic materials and permafrost within 2 m of the surface. Despite these requirements, cryosols contain five of the eight diagnostic epipedons recognized in ST, including two organic horizons (folistic and histic) and three mineral horizons (mollic, umbric, ochric).

The folistic epipedon is an organic horizon that forms under unsaturated conditions and must be more than 15 or 20 cm thick. Soils with folistic epipedon (Folistels or folistic subgroups) occur on peat plateaus or mountains in Alaska and likely occur elsewhere in the world. These soils are well-drained, have an udic soil-moisture regime, and contain a layer of mossy or woody peat over mineral material and/or permafrost. The histic epipedon is a thick organic horizon that forms under saturated conditions. Soils with a histic epipdeon are recognized in the Histels suborder (Organic cryosols) or in mineral soils. The histic epipedon in Histels must be more than 40 cm thick and comprise 80 % or more of the upper 50 cm of soil. The histic epipedon is recognized at the great-group level in Turbels and Orthels and must be at least 20 cm thick, i.e., comprise more than 40 % of the upper 50 cm. Soils with a histic epipedon are very common in the circumarctic and occur in depressions with restricted drainage in mountains throughout the world and occur to a limited extent along the western Antarctic Peninsula and its offshore islands and along the coast in East Antarctica.

The umbric epipedon is a thick, dark-colored mineral horizon that is enriched in organic matter but that contains a low base saturation. Soils with an umbric epipedon are common in the circumarctic (Tarnocai and Bockheim 2011); they are particularly common in mountains containing permafrost (Bockheim and Munroe 2014); and they occur to a limited extent on the western Antarctic Peninsula and its offshore islands (Michel et al. 2006; Simas et al. 2007). The mollic epipedon is a thick, dark-colored, highly base saturated mineral horizon that is enriched in organic matter. Soils with a mollic epipedon occur in the Kuparuk River drainage of the North Slope of Alaska (Ping et al. 1998) and occur in the high mountains where the parent materials are base-rich (Bockheim and Koerner 1997). The ochric epipedon is lighter in color, lower in organic matter, or too thin to satisfy the requirements of the mollic or umbric epipedon. The ochric epipedon is common in the High Arctic (Tarnocai and Bockheim 2011), the Antarctic mountains, and in the nival zone of high mountains (Bockheim and Munroe 2014).

Because of intensive cryoturbation, only 8 of the 19 diagnostic subsurface horizons occur in cryosols. The albic horizon is a subsurface horizon that is 1 cm or more in thickness, is light-colored, eluvial, and leached of clay and Fe oxides. Albic horizons are common in Spodosols underlain by permafrost in the Russian Arctic (Jakobsen et al. 1996; Alekseev et al. 2003), northeast Greenland (Ugolini 1966), abandoned penguin rookeries in Antarctica (Bölter et al. 1997; Beyer and Bölter 2000), and in the subalpine zone in mountains underlain by permafrost (Burns 1990; Skiba 2007). The argillic horizon features an accumulation of illuvial, high-activity silicate clays, or the presence of clay bridges or coatings. The author is aware of only two cryosol pedons with an argillic horizons reported in the literature, including one in Chersky, Russia (Smith et al. 1995) and one in the Yukon Territory (Tarnocai et al. 1991). Both soils occur on nonsorted stone polygons.

Calcic horizons are moderately thick, contain abundant secondary carbonates, and lack cementation. These occur in central Yakutia, Russia (Desyatkin et al. 2011), Svalbard (Kabala and Zapart 2012), and in alpine soils (Kann 1965; Nimlos and MConnell 1965; Knapik et al. 1973; Bockheim and Koerner 1997). Cambic horizons reflect subdued pedogenesis but must have a texture of loamy fine sand or finer. Soils with cambic horizons are common in high-mountains environments (Bockheim and Munroe 2014) and are likely common in high-latitude environments as well.

Gypsic and petrogypsic horizons occur in Antarctica (Bao et al. 2000). Natric horizons feature accelerated clay illuviation, due to the dispersive properties of Na. Soils containing these horizons are common in central Yakutia (Desyatkin et al.

2011). Soils with salic horizons (salts more soluble than gypsum) are common under anhydrous conditions in central and southern Victoria Land, Antarctica (Bockheim 1997). Spodic horizons contain illuvial organic matter and Al (with or without Fe), are dark-colored, and are low in base cations. Spodic horizons have been reported in the lower Kolyma River valley (Jakobsen et al. 1996; Alekseev et al. 2003), in northeast Greenland (Ugolini 1966), and in abandoned penguin rookeries in East Antarctica (Beyer and Bölter 2000) and King George Island (Bölter et al. 1997), Antarctica. Soils with spodic horizons are common in the subalpine zone in mountains underlain by permafrost (Burns 1990; Skiba 2007).

In addition to diagnostic horizons, cryosols contain a number of distinguishing characteristics or properties, as defined in ST. Several of these properties are unique to cryosols, including anhydrous conditions (less than 3 % moisture by weight), cryoturbation (frost mixing and sorting), gelic materials (organic or mineral materials showing cryoturbation), glacic layers (massive ice or ground ice 30 cm or more thick), permafrost (remains below 0 °C for 2 or more yr in succession), and gelic soil-temperature regimes (mean annual soil temperature at 50 cm at or below 0 °C).

There are several other important diagnostic characteristics that have been identified in cryosols. The Coefficient of Linear Extensibility (COLE) is manifested in many cryosols by the characteristic of dilatancy (Jones et al. 2010; Karavaeva 2013). MacNamara and Tedrow (1966) characterized a Grumusol (Vertisol-like soil) with permafrost near Umiat, Alaska (69°23′N, 152°10′W). The soil featured gilgai topography and desiccation cracks 2–3 cm in width and 8–20 cm in depth. The soil featured salt segregations on ped surface, a strongly granular structure, abundant water-soluble Na and SO_4 (thenardite), and clay concentrations ranging between 76 and 82 % that were dominantly smectites.

Volcanic glass is present in permafrost-affected soils of the Kamchatka Peninsula, Russia and in soils on Deception Island in the South Shetland Group off the Antarctic Peninsula. However, in both cases the glass has not weathered sufficiently to yield andic soil properties.

Cryosols either have an aquic or a udic soil-moisture regime. cryosols in maritime Antarctica, especially on Seymour Island contain sulfidic materials and a sulfuric horizon (de Souza et al. 2012).

6.4 Summary

Soil classification schemes have moved from genetic or zonal systems to natural or technical systems with the publication of the Seventh Approximation (1960). Soils underlain by permafrost exist in a separate category at the highest level in the Canadian and WRB systems (cryosols) and *Soil Taxonomy* (ST) (Gelisols). In ST the primary criteria are the occurrence of gelic materials, which are organic or mineral materials that are subjective to cryoturbation, cryodesiccation, and ice segregation, as well as the existence of permafrost within 1–2 m of the ground surface. Other diagnostic horizons are found in Gelisols/cryosols but these play a lesser role than the presence of gelic materials.

References

Alekseev A, Alekseeva T, Ostroumov V, Siegert C, Gradusov B (2003) Mineral transformations in permafrost-affected soils, north Kolyma lowland. Russ Soil Sci Soc Am J 67:596–605

Baldwin M, Kellogg CE, Thorp J (1938) Soil classification. Soils and men. U.S. Department of Agriculture, Yearbook of the U.S. Government Printing Office, Washington, DC, pp 979–1001

Bao H, Campbell DA, Bockheim JG, Thiemens MH (2000) Origins of sulphate in Antarctic dry-valley soils as deduced from anomaloug 17O compositions. Nature 407:499–502

Beyer L, Bölter M (2000) Chemical and biological properties, formation, occurrence and classification of Spodic Cryosols in a terrestrial ecosystem of East Antarctica (Wilkes Land). Catena 39:95–119

Bockheim JG (1997) Properties and classification of Cold Desert soils from Antarctica. Soil Sci Soc Am J 61:224–231

Bockheim JG, Koerner D (1997) Pedogenesis in alpine ecosystems of the High Uinta Mountains, Utah. Arctic Alp Res 29:164–172

Bockheim JG, Munroe JS (2014) Organic carbon pools and genesis of alpine soils with permafrost: a review. Arctic Antarct Alp Res 46

Bockheim JG, Ugolini FC (1990) A review of pedogenic zonation in well drained soils of the southern circumpolar regions. Quat Res 34:47–66

Bölter M, Blume H-P, Schneider D, Beyer L (1997) Soil properties and distributions of invertebrates and bacteria from King George Island (Arctowski Station), maritime Antarctic. Polar Biol 18:295–304

Bryan WH, Teakle LJH (1949) Pedogenic inertia: a concept in soil science. Nature 164(3):969

Burns, SF (1990) Alpine Spodosols: Cryaquods, Cryohumods, Cryorthods, and Placaquods above treeline. In: Kimble JM, Yeck RD (eds) Proceedings of the 5th International Correlation Meeting (ISCOM): characterization, classification, and utilization of Spodosols. USDA Soil Conservation Service, Lincoln, pp 46–62

Campbell IB, Claridge GGC (1969) A classification of frigic soils – the zonal soils of the Antarctic continent. Soil Sci 107:75–85

De Souza JJII, Schaefer CEGR, Abrahão WAP, de Mello JWV, Simas FNB, da Silva J, Francelino MR (2012) Hydrogeochemistry of sulfate-affected landscapes in Keller Peninsula, maritime Antarctica. Geomorphology 155–156:55–61

Desyatkin RV, Lesovaya SN, Okoneshnikov MV, Zaitseva TS (2011) Palevye (pale) soils of central Yakutia: genetic specificity, properties, and classification. Eurasian Soil Sci 88:1304–1314

IUSS Working Group WRB (2006) World reference base for soil resources 2006, 2nd edn, World soil resources report, no. 103, FAO, Rome

Jakobsen BH, Siegert C, Ostroumov V (1996) Effect of permafrost and palaeo-environmental history on soil formation in the lower Kolyma lowland, Siberia. Geografisk Tidsskrift 96:40–50

Jones A, Stolbovoy V, Tarnocai C, Broll G, Spaargaren O, Montanarella L (eds) (2010) Soil atlas of the Northern circumpolar region. European Commission, Publications Office of the European Union, Luxembourg

Kabala C, Zapart J (2012) Initial soil development and carbon accumulation on moraines of the rapidly retreating Werenskiold Glacier, SW Spitsbergen, Svalbard archipelago. Geoderma 175–176:9–20

Kann LA (1965) High-mountain soils of the western Pamirs. Soviet Soil Sci (9):1028–1036

Karavaeva NA (2013) Soil zonality of the Chukotka upland. Eurasian Soil Sci 46:468–483

Knapik LJ, Scotter GW, Pettapiece WW (1973) Alpine soil and plant community relationships of the Sunshine Area, Banff National Park. Arctic Alp Res 5:A161–A170

MacNamara EE, Tedrow JCF (1966) An Arctic equivalent of the Grumusol. Arctic 19(2):145–152

Mazhitova G (2004) Classification of Cryosols in Russia. In: Kimble JM (ed) Cryosols: permafrost-affected soils. Springer, New York, pp 611–626

Michel RFM, Schaefer CEGR, Dias LE, Simas FNB, Benites VM, Mendonça ES (2006) Ornithogenic Gelisols (Cryosols) from maritime Antarctica: pedogenesis, vegetation, and carbon studies. Soil Sci Soc Am J 70:1370–1376

Nimlos TJ, McConnell RC (1965) Alpine soils in Montana. Soil Sci 99:310–321

Ping C-L, Bockheim JG, Kimble JM, Michaelson GJ, Walker DA (1998) Characteristics of cryogenic soils along a latitudinal transect in Arctic Alaska. J Geophys Res 103:917–928

Simas FNB, Schaefer CEGR, Melo VF, Albuquerque-Filho MR, Michel RFM, Pereira VV, Gomes MRM, da Costa LM (2007) Ornithogenic cryosols from maritime Antarctica: phosphatization as a soil forming process. Gedoerma 138:191–203

Skiba M (2007) Clay mineral formation during podzolization in an alpine environment of the Tatra Mountains, Poland. Clays Clay Miner 55:618–634

Smith CAS, Swanson DK, Moore JP, Ahrens RJ, Bockheim JG, Kimble JM, Mazhitova GG, Ping CL, Tarnocai C (1995) A description and classification of soils and landscapes of the lower Kolyma River, northeastern Russia. Polar Geogr Geol 19:107–126

Soil Classification Working Group (1998) The Canadian system of soil classification, 3rd edn, Research Branch, Agriculture & Agri-Food Canada, Publication, 1646, NRC Press, Ottawa

Soil Survey Staff (1960) Soil classification, a comprehensive system, 7th approximation. U.S. Government Printing Office, Washington, DC

Soil Survey Staff (1975) Soil taxonomy: a basic system of soil classification for making and interpreting soil surveys, U.S. Department of Agriculture, Soil Conservation Service, Agriculture handbook no. 436. Superintendent of Documents of the U.S. Government Printing Office, Washington, DC

Soil Survey Staff (1999) Soil taxonomy: a basic system of soil classification for making and interpreting soil surveys, 2nd edn, U.S. Department of Agriculture, Soil Conservation Service, Agriculture handbook no. 436. Superintendent of Documents of the U.S. Government Printing Office, Washington, DC

Soil Survey Staff (2010) Keys to soil taxonomy, 11th edn. USDA/Natural Resources Conservation Service, Washington, DC

Tarnocai C, Bockheim JG (2011) Cryosolic soils of Canada: genesis, distribution, and classification. Can J Soil Sci 91:749–762

Tarnocai C, Broll G, Blume H-P (2004) Classification of permafrost-affected soils in the WRB. In: Kimble JM (ed) Cryosols: permafrost-affected soils. Springer, New York, pp 637–656

Tarnocai C, Kroetsch D, Gould J, Veldhuis H (1991) Soils, vegetation and trafficability, Tanquary Fiord and Lake Hazen areas, Report RM911/ELS. Ellesmere Island National Park Reserve, Canadian Parks Service, Winnipeg, 269 pp

Tedrow JCF (1968) Pedogenic gradients of the polar regions. J Soil Sci 19:197–204

Tedrow JCF (1977) Soils of the polar landscapes. Rutgers University Press, New Brunswick, New Jersey

Tedrow, JCF (1991) Pedologic linkage between the cold deserts of Antarctica and the polar deserts of the high Arctic. In: Contributions to Antarctic research II. AGU. Antarct Res Ser 53:1–17

Ugolini FC (1966) Soils of the Mesters Vig district, northeast Greenland. Meddelelser om Grønland 176:1–22

Ugolini FC (1986) Pedogenic zonation in the well-drained soils of the Arctic regions. Quat Res 26:100–120

Chapter 7
Distribution of Cryosols

7.1 Introduction

According to the World Reference Base for Soil Resources (IUSS WRB Working Group 2006) and the International Permafrost Association (IPA) cryosol working group, cryosols cover an area of 18 million km^2 (13 % of ice-free land area). By contrast *Soil Taxonomy* (Soil Survey Staff 1999) reports that gelisols cover an area of 11.3 million km^2 (8.6 % of ice-free land area). According to the IPA website, permafrost covers about 23 million km^2 or 25 % of the ice-free land area in the Northern Hemisphere only (http://ipa.arcticportal.org). Recent estimates by Gruber (2012) from a high-resolution global model suggest that the permafrost area is 22 ± 3 million km^2.

In that some of the permafrost occurs below 1–2 m, the distribution of cryosols estimated here may be 13.8 million km^2 (Table 7.1). The estimates in Table 7.1 suggest that 63 % of the permafrost zone contains cryosols, because of deep active layers especially on the Qinghai-Tibet Plateau and in Canada. About 83 % of the cryosols occur in the circumarctic, 17 % in high-mountain regions, primarily at the high latitudes, and only 0.1 % in Antarctica.

7.2 Distribution of Cryosols

7.2.1 *Circumarctic Lowlands and Hills*

The majority of the cryosols in the circumarctic lowlands and hills are in Russia (9.7 million km^2) and Canada (2.5 million km^2), with smaller areas in Alaska (793,000 km^2), China (320 million km^2), and Greenland (250,000 km^2) (Table 7.1). According to the Soil Survey Staff (1999), cryosols in the circumarctic are distributed: Turbels (56 %), Orthels (36 %), and Histels (9 %). Cryosols in Canada

J.G. Bockheim, *Cryopedology*, Progress in Soil Science,
DOI 10.1007/978-3-319-08485-5_7

Table 7.1 World distribution of cryosols

Region	Area (10^3 km^2)	% of total	References
Russia	9,720	70.6	Jones et al. (2010)
Canada	2,500	18.2	Tarnocai and Bockheim (2011)
USA	793	5.8	Soil Survey Staff (1999)
China	320	2.3	
Greenland	250	1.8	Jones et al. (2010)
Mongolia	64	0.5	Maximovich (2004)
Antarctica	46	0.3	
Svalbard	24	0.2	Jones et al. (2010)
Kyrgyzstan	11	0.1	
Tajikistan	8	0.1	
Others	23	0.2	
Total	13,759	100	

(Tarnocai and Bockheim 2011) can be ranked Turbic cryosols (72 %) Organic cryosols (16 %), and Static cryosols (12 %) (Fig. 7.1).

7.2.2 *Antarctica*

Cryosols only account for 49,500 km^2 in Antarctica, 0.35 % of the continent. The largest ice-free areas containing cryosols include the Transantarctic Mountains of Victoria Land (24,200 km^2; 49 % of total), Palmer and Graham Land of the Antarctic Peninsula (10,000 km^2; 20 %), and MacRobertson Land in East Antarctica (5,400 km^2; 11 %). The ranking of soils by suborder for ice-free areas of Antarctica is Orthels (44 %), Turbels (36 %), and Histels (0.4 %) (Table 9.6). Approximately 16 % of the ice-free areas contain soils other than Gelisols, especially Gelorthents and Gelepts.

7.2.3 *Alpine Permafrost Regions*

Soils in the high mountains comprise 17 % of the global area of cryosols (Table 7.1). The main regions featuring cryosols include high-latitude mountain ranges such as the Ural and Sayan Mountains of Russia, the Tien Shan Mountains and Qilian Mountains of China, the Brooks Range and Wrangel Mountains of Alaska, the Canadian Cordillera, and the Altai Mountains of Mongolia. These areas contain primarily Orthels (Aquorthels, Haplorthels, Historthels, Mollorthels, and Umbrorthels), some Turbels (Aquiturbels, Histoturbels) and a lesser proportion of Histels (Fibristels).

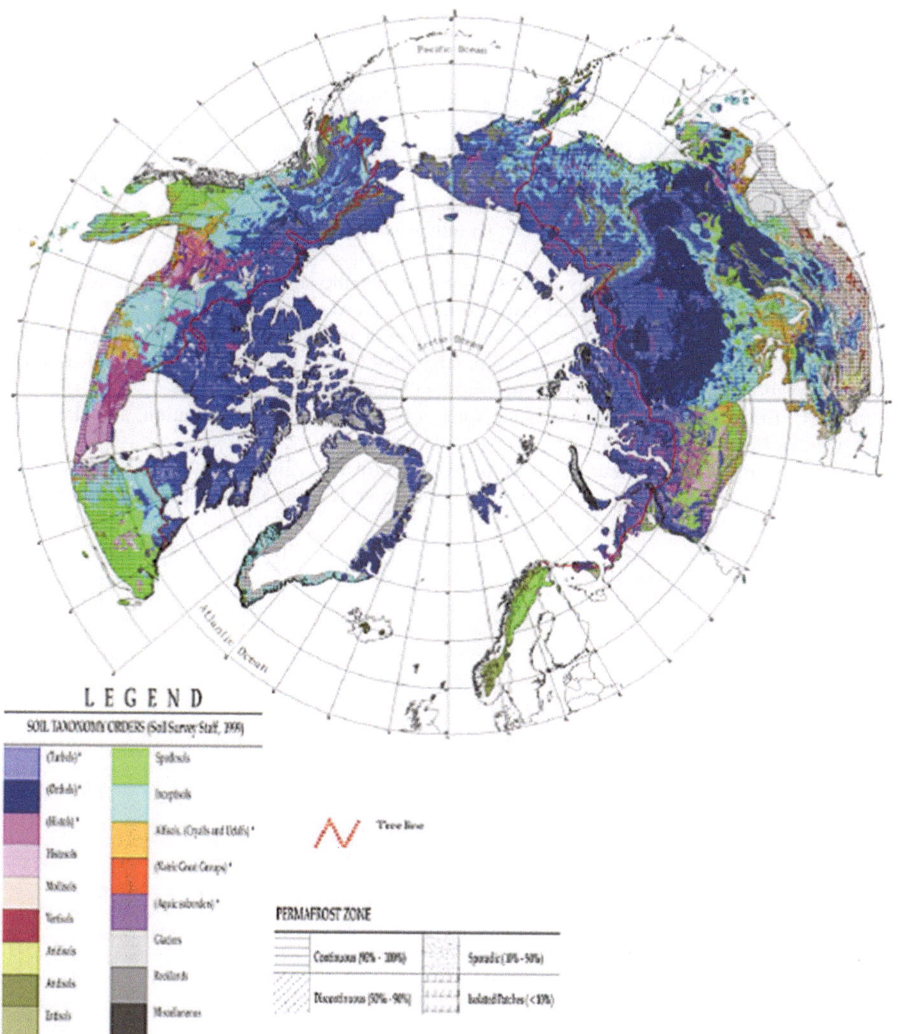

Fig. 7.1 Distribution of soils I the northern circumpolar region based on the NCSCD (Tarnocai et al. 2009)

7.3 Summary

Cryosols cover about 13.8 million km² worldwide, of which 83 % occur in the circumarctic and 17 % in mountains of the high latitudes and high mountains. Because the active layer is too deep (>1 or 2 m) in mountains with permafrost and in the sporadic and isolated permafrost zones of the circumarctic, only 63 % of the permafrost region contains cryosols. The other soils are Entisols, Inceptisols, Histosols, and other soil orders.

References

Gruber S (2012) Derivation and analysis of a high-resolution estimate of global permafrost zonation. The Cryosphere 6:221–233

IUSS Working Group WRB (2006) World reference base for soil resources, World soil resources reports no. 103. FAO, Rome

Jones A, Stolbovoy V, Tarnocai C, Broll G, Spaargaran O, Montanarella (2010) Soil atlas of the northern circumpolar region. European Commission, Office for Official Publications of EC, Luxembourg, 142 pp

Maximovich SV (2004) Geography and ecology of cryogenic soils of Mongolia. In: Kimble JM (ed) Cryosols: permafrost-affected soils. Springer, New York, pp 253–274

Soil Survey Staff (1999) Soil taxonomy: a basic system of soil classification for making and interpreting soil surveys, 2nd edn, Agriculture handbook no. 436. USDA, Natural Resources Conservation Service, US Government Printing Office, Washington, DC

Tarnocai C, Bockheim JG (2011) Cryosolic soils of Canada: genesis, distribution, and classification. Can J Soil Sci 91(5):749–762. doi:10.4141/CJSS10020

Tarnocai C, Canadell JG, Schuur EAG, Kuhry P, Mazhitova G, Zimov S (2009) Soil organic carbon pools in the northern circumpolar permafrost region. Global Biogeochem Cycles 23:GB2023. doi:10.1029/2008GB003327

Chapter 8
Cryosols of the Circumarctic Region

8.1 Introduction

As seen in the previous chapter, the circumarctic region contains 11.4 million km^2, or 83 % of the cryosols worldwide. These soils are divided on the basis of various physiographic schemes developed by country. The Russian Arctic is commonly divided into the Arctic archipelagos, northeastern Eurasia, the European North, western Siberia, and central Siberia (Kimble 2004). Canada is often divided into the High Arctic, the Mid-Arctic, and the Low Arctic (Tarnocai 2004). Arctic Alaska contains the Arctic Coastal Plain, the Arctic Foothills, western Alaska, and interior Alaska (Ping et al. 2004). Cryosols fringe the coast in Greenland from about 69°N to the tip at Cape Morris Jesup at 83°38′S). Soils of the mountains within the circumarctic region are considered in Chap. 10.

8.2 Soil-Forming Factors

8.2.1 Climate, Permafrost, and Active-Layer Thickness

The climate of the circumarctic is determined by proximity to oceans, elevation, and latitude. The mean annual air temperature ranges from −0.5° to about −8 °C in the Low Arctic of Eurasia, North America and Greenland, from −8 to −16 °C in the Mid-Arctic, and from −16 to −18 °C in the High Arctic (Table 8.1). The mean temperature of July, the warmest month, ranges from 5.7 to 17 °C in the Low Arctic, from 4.5 to 12 °C in the Mid-Arctic, and from 1 to 7.6 °C in the High Arctic. The mean annual precipitation ranges from 150 to 1,200 mm in the Low Arctic, from 100 to 400 mm in the Mid-Arctic, and from 100 to 200 mm in the High Arctic.

Most of the high-latitude areas with cryosols have continuous permafrost, but discontinuous permafrost is present in the Russian European North and central

J.G. Bockheim, *Cryopedology*, Progress in Soil Science,
DOI 10.1007/978-3-319-08485-5_8

Table 8.1 Climate of the circumarctic

Location	Lat (°N)	Long (°)	Elevation (m)	MAAT (°C)	MJulyT (°C)	MAP (mm)	Permafrost[a]	ALD (m)
Russia								
Arctic Archipelagos	71–81	52–180E		−1.5 to −10	1.0–6.0	125–300	C	0.3–2.0
Northeastern Eurasia	60–73	140E–170W		−4.1 to −16	4.0–12	100–600	C	0.4–2.0
European North	67–70	30–65E		−0.8 to −7.6	5.8–12	270–800	C, D	0.3–5.0
Western Siberia	62–73	65–100E		−3.4 to −11	5.7–17	115–400	C	0.3–1.0
Central Siberia	54–78	80–135E		−0.5 to −16	11–19	150–1,000	C, D	0.2–2.5
Canada								
Low-Arctic	58–69	141–63W		−8.2 to −14	8.9–11	150–200	C	0.69–1.3
Mid-Arctic	63–75	125–65W		−15 to −16	4.5–4.9	100–250	C	0.60–0.85
High-Arctic	67–84	125–67W		−17 to −18	2.0–4.0	100–175	C	0.25–0.85
Alaska								
Arctic Coastal Plain	70–72	145–163W		−10 to −13	4.9–7.6	125–150	C	0.45–1.0
Arctic Foothills	68–70	142–163W		−8.3 to −8.8		140–270	C	0.45–0.8
Western Alaska lowlands	64–67	160–167W		−0.5 to −4.0	11–15	250–500	D	0.2–2.0
Interior Alaska	64–66	140–167W		−2.5 to −6.0	15–17	230–340	D	0.2–2.0
Greenland								
Southern	60–68	20–55W		−1.3 to −2.0	6.4–6.7	775–980	C	0.4–0.9
Central	68–76	19–75W		−4.4 to −7.2	5.2	250–275	C	0.4–0.9
Northern	76–83	12–67W		−12 to −17	3.7	140–200	C	0.4–0.9
Fennoscandia	69–71	18–30W		−3.9 to −6.1	12–13	450–1,200	D	1.4–3.0

[a]Permafrost distribution: *C* continuous (70–90 %), *D* discontinuous (50–70 %)

Siberia, the western Alaska lowlands and interior Alaska, and parts of the Low Arctic of Canada. Active-layer depths range from 0.4 to 2.0 m in the Low Arctic, from 0.4 to 1.0 m in the Mid-Arctic, and from 0.25 to 0.9 m in the High Arctic (Table 8.1).

8.2.2 Biota

Based on the circumarctic vegetation map of Walker et al. (2005) (see Fig. 4.2), the dominant vegetation types are erect shrub land (25 %), peaty graminoid tundra (18 %), barrens (12 %), mineral graminoid tundra (11 %), prostrate-shrub tundra (11 %), and wetlands (7 %). Examples of these vegetation types are given in Fig. 8.1.

Migrating birds congregate in the circumarctic, especially in wetlands and in the coastal areas. These birds play an important role locally in pedogenesis. For example, Zwolicki et al. (2013) showed the effects of guano deposition and nutrient enrichment in seabird colonies of Spitsbergen. A recent Arctic biodiversity

Fig. 8.1 Common vegetation types in the circumarctic: (**a**) Wet graminoid tundra, Low-Arctic, Arctic Coastal Plain, Alaska. (**b**) Cryptogam herb barren, High Arctic, northern Greenland. (**c**) Erect dwarf-shrub tundra. (**d**) Low-shrub tundra, Low-Arctic, Seward Peninsula, Alaska. (**e**) Moist non-tussock sedge, dwarf-shrub, nonacidic tundra, Arctic Foothills, Alaska. (**f**) Prostrate dwarf-shrub herb tundra, High Arctic, Axel Heiberg Island, Nunavut, Canada. (**g**) Tussock sedge dwarf-shrub tundra, Arctic Foothills, Alaska (All photos compliments of D. Walker, Alaska Geobotany Center, University of Alaska-Fairbanks; http://www.arcticatlas.org)

assessment (Lovejoy 2013) showed that bacteria, archaea and single-celled eukaryota (protists) are ubiquitous and diverse members of biological communities in Arctic soils. Humans have had a major impact of terrestrial ecosystems in the circumarctic; this topic will be considered in Chaps. 12 and 13.

8.2.3 *Relief*

The relief in the arctic archipelagos, coastal plains, lowlands and along the river terraces is flat to undulating with elevations generally below 200 m. In the Arctic Foothills and inland plateaus the relief is steeper but the elevations are commonly below 500 m. The mountains have even steeper relief and higher elevations, such as the Brooks Range of Alaska, the Rocky Mountains of Alaska and Canada, the Richardson Mountains of Canada, and the Yablonai-Sayan-Stanovoi Mountains of Siberia. However, these areas are considered in Chap. 10.

8.2.4 *Parent Materials and Time*

Cryosols in the circumarctic are derived from a variety of parent materials, including marine and lacustrine sediments in the northern Eurasian and North American coastal plains, loess (yedoma in central Siberia and the Russian Far East) and northern Alaska, glacial deposits, alluvium along the large arctic rivers (such as the Ob, Yenisey, Lena, and Kolyma rivers in Russia, the Yukon in Alaska and Canada, and Mackenzie river in Canada), and sand dunes and sand plains in the Arctic Coastal Plain (Table 8.2). Volcanic ejecta are an importance parent material in arctic portions of the circumpacific belt. Peat deposits are common throughout the circumarctic, especially in the Mackenzie River delta. Most of these materials are of Last Glacial Maximum (ca. 12–20 ky BP) age or younger. In the Arctic Coastal Plain the thaw-lake cycle has interrupted soil formation since at least the last glaciation (Hinkel et al. 2003).

Table 8.2 Common parent materials in the circumarctic lowlands and hills

Zone	Russia	Canada	Alaska	Greenland
High Arctic	Marine, alluvium, sand dunes & plains	Lacustrine, alluvium, colluvium, marine	[not present]	Glacial, colluvium
Mid-Arctic	Loess, glacial, alluvium, colluvium, volcanic ash	Lacustrine, alluvium, sand dunes & plains	Marine, lacustrine, alluvium, sand dunes & plains, loess	Glacial, alluvium
Low Arctic	Alluvium, peat	Peat, lacustrine, alluvium	Loess, glacial, peat, lacustrine, volcanic ash	Glacial, alluvium,

8.3 Soil Properties

As can be expected, with the wide variation in soil-forming factors across the circumarctic, soil properties vary accordingly. Key morphological features of arctic soils are the presence of permafrost within 1 or 2 m of the surface, cryoturbation and cryodesiccation within the active layer, and accumulation of segregation ice in the transient layer and near-surface permafrost. Representative profiles from arctic Russia, Canada, and Alaska are shown in Figs. 8.2, 8.3, and 8.4, respectively. Cryoturbation is readily apparent in Figs. 8.2a–c, 8.3c, 8.4a, and 8.5. Segregated ice can be seen in Figs. 8.3b and 8.4b. The presence of a shallow active layer often results in redoximorphic conditions (Fig. 8.2d). Figure 8.2a shows a silt-enriched horizon that has dilatancy, and Fig. 8.3a shows a desert pavement and weak cryodesiccation.

Many soils of the circumarctic are neutral to slightly alkaline (Table 8.3); however, organic horizons and sandy podzolized soils may be extremely acid. These trends are reflected in the degree of base saturation. In general arctic soils have abundant organic C in the active-layer and transition layer. Because of the high levels of SOC and occasionally also of clay, cation-exchange values commonly are high in arctic soils.

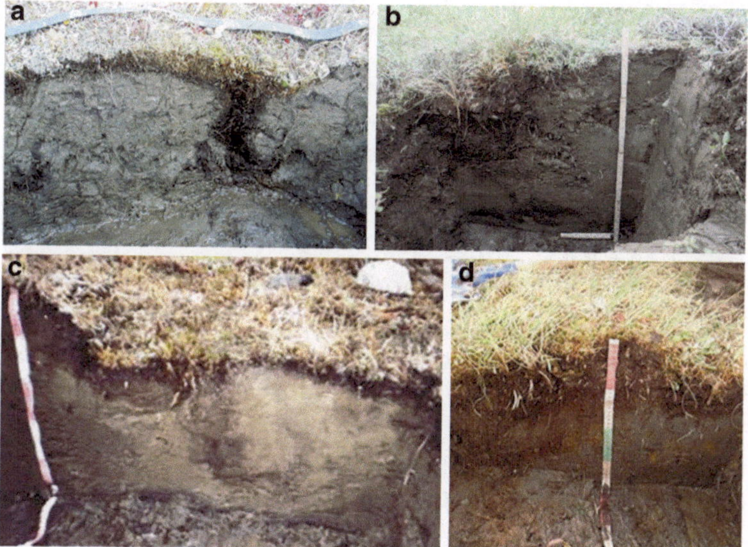

Fig. 8.2 Representative soils in the Russian Arctic: (**a**) Aquiturbel, northern Russia (Joint Research Commission, EU). (**b**) Ruptic-Histic Aquiturbel, Cherskiy, Chutkotka (G. Hugelius). (**c**) Aquiturbel, Cape Chukotskii (http://www.arcticportal.org). (**d**) Histoturbel (Peaty Gleyzem), Tiksi, Russia (J. Antsibor)

Fig. 8.3 Representative soils in the Canadian Arctic: (**a**) Static cryosols, (**b**) Organic cryosols, and (**c**) Turbic cryosols. Note the ice (*white areas*) in the lower part of soil profiles **b** and **c** (Tarnocai and Bockheim 2011)

8.4 Soil-Forming Processes

Cryoturbation is the dominant soil-forming process in the circumarctic. Additional processes of importance are biological enrichment of base cations, paludization, podzolization, gleization, melanization, cambisolization, and base cation leaching. Processes such as argilluviation, andisolization,

Fig. 8.4 Representative soils in the Alaskan Arctic lowlands and hills: Haploturbels (*top*), Haplorthels (*lower, left*), and Sapristels (*lower, right*) (Bockheim and Tarnocai 2012)

salinization, solonization, solodization, and vertization function to a limited extent and in restricted areas. Figure 8.5 shows gradients in pedogenic processes northward from the forest-tundra across the Low-Arctic and Subarctic Tundra, the Mid-Arctic Tundra, and the High Arctic Barrens (Goryachkin et al. 2004). The figure shows a reduction in redoximorphism (gleization), melanization (organic matter accumulation), podzolization, and textural differentiation

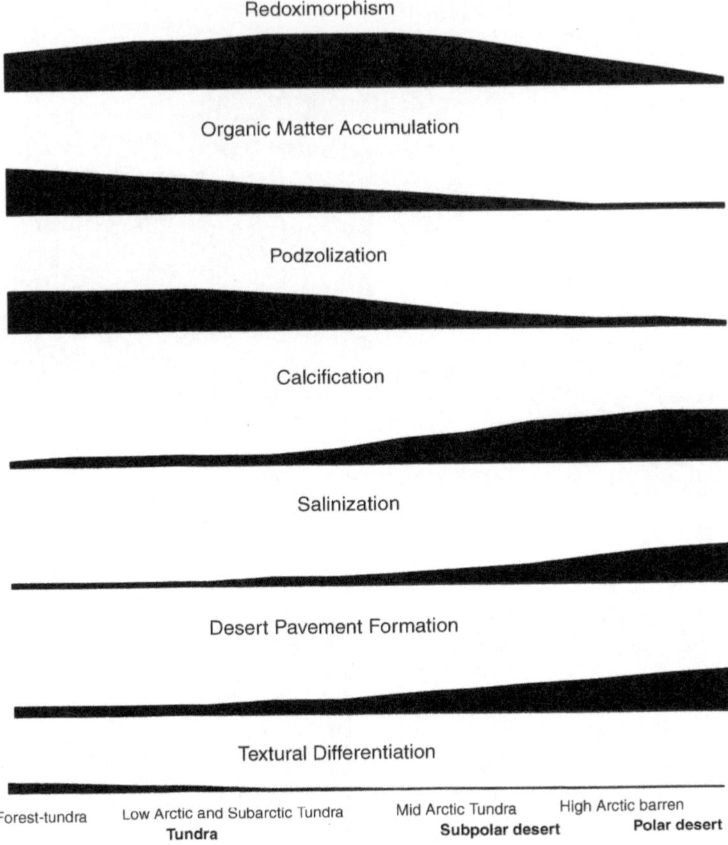

Fig. 8.5 Latitudinal gradient in pedogenic processes in the circumarctic (Goryachkin et al. 2004)

and an increase in latitude and an increase in calcification, salinization, and desert pavement formation as the climate becomes more arid.

8.5 Soil Classification and Geography

Figure 7.1 (previous chapter) shows the distribution of soils in the northern circumpolar region. Based on this map and the databases from which it was derived, Turbels are the dominant soils in the circumarctic permafrost region, covering approximately $6{,}454 \times 10^3$ km^2, followed by Histels ($2{,}714 \times 10^3$ km^2), and Orthels ($1{,}051 \times 10^3$ km^2). The countries with the largest area of cryosols (excluding mountain cryosols) are Russia (8.2 million km^2), Canada (2.3 million km^2), the USA (503,000 km^2), and Greenland (250,000 km^2).

Table 8.3 Analytical properties of Arctic cryosols

Site	Horizon	Depth (cm)	pH H_2O	OC (%)	CEC ($cmol_c/kg$)	BS (%)	Silt (%)	Clay (%)	$CaCO_3$ (%)
3	Typic Historthels; Prudhoe Bay, AK (Ping et al. 1998)								
	Oi	0–21	7.5	23	59	100			
	Oa/Oi	21–39	7.1	18	54	100	53	20	
	Oajj/Cgf	39–67	7.1	14	42	100	65	19	
6	Mollic Aquiturbels; Sagwon Hills, AK (Ping et al. 1998)								
	Oi	0–9	7.4	36	103	95			
	Oejj1	9–21	7.2	24	102	90			
	Bwjj	21–47	7.0	4	22	100	70	20	0.1
	Bg	47–56	7.2	4	24	100	70	23	0.1
	Oejj2	56–74	6.9	20	98	88			1
	2Ajj	74–86	6.8	7	30	100	69	17	
	2Cg/Oejjf	86–100		11	51	100	70	18	3
5	Typic Umbriturbels, Parry Peninsula, NWT, Canada (Tarnocai and Bockheim 2011)								
	Bwjj	0–19	6.1	6.7			30	28	40
	BCkjj	19–40	7.6	6.5			29	28	48
	Ckjj	40–69	7.5	6.2			19	30	52
	2Ck	69–110	7.4	6.4			31	52	51
	2Ckf	110–120	7.4	6.3			31	52	51
2	Typic Molliturbels, Prudhoe Bay, AK (Ping et al. 1998)								
	A	0–22	7.3	21	90	90	48	16	
	Oa/A	0–24	7.4	15	62	84	32	13	3
	O'i	22–42	6.7	26	72	75			
	2 AC	0–42	6.6	7	26	73	10	4	5
	2Oejj/Cgf	43–50	6.5	10	29	76	17	4	10
1	Typic Psammoturbels, Ellesmere Island, Nunavut, Canada (Tarnocai and Bockheim 2011)								
	Ckjj1	0–19	6.6	0.97			11	6.2	5.1
	Ckjj2	19–44	6.7	0.2			27	9.1	3.1
	Ckjj3	44–65	6.7	0.26			11	4.8	1.4
	Ckjjf	65–85	6.7	0.2			20	7.1	1.4
1066	"Oxyaquic Natrothels", central Yakutia, Russia (Sokolov et al. 2004)								
	O	2–6	4.4		50				
	AO	6–9	4.5	34	48		40	22	
	Eg	9–23	6.2	1.3	32		32	10	
	Btn	23–34	8.0	2	51		51	33	
	Bk	50–65	8.7	1.2	38		38	21	
10	Typic Aquorthels, Imnavait Creek, AK (Ping et al. 1998)								
	Oi	0–7	4.6	47	103	28			
	Oe1	7–13	4.9	44	70	25			
	Oe2	13–18	4.9	25	58	17			
	Bw	18–30	4.9	7	22	11	40	14	

(continued)

Table 8.3 (continued)

Site	Horizon	Depth (cm)	pH H_2O	OC (%)	CEC ($cmol_c$/kg)	BS (%)	Silt (%)	Clay (%)	$CaCO_3$ (%)
	Bgjj	30–35	5.6	5	17	10	37	18	
	Cgf1	35–62	6.5	4	12	13	32	16	
	Cgf2	62–82	6.5	3	13	20	38	15	
	Typic Mollorthels, Novaya Zemlya, Russia (Goryachkin and Karavaeva 2004)								
	Ak1	0–10	7.7	10					6
	Ak2	10–20	8.3	6.8					5.5
	Bk1	20–28	8.6	1.3					2.1
	Bk2	28–55	8.4	0.77					0.27
	Typic Spodorthels, Novaya Zemlya, Russia (Goryachkin and Karavaeva 2004)								
	O	0–2	6.3		56	92			
	AO	2–4	5.6	9.7	25	95			
	EA	4–6	5.8	3.8	7.3	96			
	Bhs	6–16	5.5	2.4	2.3	73			
	BC	16–25	5.4	1.6	2.2	71			
	R	25–50							
1	Sapric Glacistels, Prudhoe Bay, AK (Ping et al. 1998)								
	Oa1	0–18	7.1	23	79	100			17
	Oa2	18–39	6.6	15	29	61	47	11	2
	Oaf	50–80	6.6	22	49	99	42	7	
	2Cf	39–80	8.1	0.3	2	100	10	1	22
	2Wf	50–100							
7	Typic Fibristels, Wrigley, NTW, Canada (Tarnocai 1973)								
	Oi	0–23	2.7	62	83	12			0
	Oif1	23–79	3.0	62	106	31			0
	Oif2	79–140	4.2	64	97	49			0
	Oif3	140–200	4.4	57	111	60			0
	2Cgf	200–257	6.9		39	100	47	33	0
10	Typic Hemistels, Fort Simpson, NWT, Canada (Tarnocai and Bockheim 2011)								
	Oe1	0–54	5.6	37					0
	Oe2	54–60	4.8	32					0
	Oef	60–165	6.0	31					0
	Cf	165–175	7.1	0.9			41	10	1.7

8.6 Summary

The circumarctic region contains 11.4 million km^2, or 83 % of the cryosols worldwide. The climate of the circumarctic varies geographically and is determined by proximity to oceans, elevation, and latitude. Most of the circumarctic contains continuous permafrost. However, discontinuous, sporadic, and isolated permafrost may occur in the Low Arctic. Active-layer depths range from 0.4 to

2.0 m in the Low Arctic, from 0.4 to 1.0 m in the Mid-Arctic, and from 0.25 to 0.9 m in the High Arctic. Common vegetation types in the Arctic are erect shrub land, peaty graminoid tundra, barrens, mineral graminoid tundra, prostrate-shrub tundra, and wetlands. The relief in the circumarctic commonly is flat to undulating with elevations generally below 500 m. Patterned ground is ubiquitous in the Arctic. Cryosols in the circumarctic have been derived from a variety of parent materials, including marine and lacustrine, glacial, windblown, colluvium, and residuum.

Common soil properties of Arctic soils are permafrost within 1–2 m of the surface, cryoturbation, cryodesiccation, and accumulation of segregated ice. Chemical and physical properties vary significantly in response to the action of the soil-forming factors, but the accumulation of organic matter and the development of redoximorphic features are common.

There is a strong latitudinal gradient in soil-forming processes in the Arctic. As one progresses to the north, there is a reduction in redoximorphism (gleization), melanization (organic matter accumulation), podzolization, and textural differentiation and an increase in latitude and an increase in calcification, salinization, and desert pavement formation as the climate becomes more arid.

Turbels are the dominant soils in the circumarctic permafrost region, followed by Histels and Orthels. Countries with the largest area of cryosols (excluding mountain cryosols) are Russia, Canada, the USA, and Greenland.

References

Bockheim JG, Tarnocai C (2012) Gelisols. In: Huang PM, Li Y, Sumner ME (eds) Handbook of soil sciences, 2nd edn. CRC Press, Boca Raton, pp 33–72, 82

Goryachkin SV, Karavaeva NA (2004) Cryosols in the Russian arctic archipelagos. In: Kimble JM (ed) Cryosols: permafrost-affected soils. Springer, New York, pp 139–160

Goryachkin SV, Blume H-P, Beyer L, Campbell I, Claridge G, Bockheim JG, Karavaeva NA, Targulian V, Tarnocai C (2004) Similarities and differences in Arctic and Antarctic soil zones. In: Kimble JM (ed) Cryosols: permafrost-affected soils. Springer, New York, pp 49–70

Hinkel KM, Eisner WR, Bockheim JG, Nelson FE, Peterson KM, Dai X (2003) Spatial extent, age and carbon stocks in drained thaw lake basins on the Barrow Peninsula, Alaska. Arctic Antarct Alp Res 35:291–300

Kimble JM (ed) (2004) Cryosols: permafrost-affected soils. Springer, New York, 726 pp

Lovejoy C (2013) Microorganisms. In: Arctic biodiversity assessment; status and trends in Arctic biodiversity. Conservation of Arctic Flora & Fauna, Akureyri, pp 373–382

Ping CL, Bockheim JG, Kimble JM, Michaelson GJ, Walker DA (1998) Characteristics of cryogenic soils along a latitudinal transect in arctic Alaska. J Geophys Res 103(D22):28917–28928

Ping C-L, Clark MH, Swanson DK (2004) cryosols in Alaska. In: Kimble JM (ed) Cryosols: permafrost-affected soils. Springer, New York, pp 71–94

Sokolov IA, Ananko TV, Konyushkov DY (2004) The soil cover of central Siberia. In: Kimble JM (ed) Cryosols: permafrost-affected soils. Springer, New York, pp 303–338

Tarnocai C (1973) Soils of the Mackenzie River area. Task Force on Northern Oil Development report, Information Canada, Ottawa, 73–26

Tarnocai C (2004) cryosols of Arctic Canada. In: Kimble JM (ed) Cryosols: permafrost-affected soils. Springer, New York, pp 95–117
Tarnocai C, Bockheim JG (2011) Cryosolic soils of Canada: genesis, distribution, and classification. Can J Soil Sci 91:749–762
Walker DA, Raynolds MK, Daniels FJA, Einarsson E, Elvebakk A, Gould WA, Katenin AE, Kholod SS, Markon CJ, Melnikov ES, Moskalenko NG, Talbot SS, Yurtsev BA (2005) The circumpolar arctic vegetation map. J Veg Sci 16:267–282
Zwolicki A, Zmudczynska-Skarbek MZ, Iliszko L, Stempniewicz L (2013) Guano deposition and nutrient enrichment in the vicinity of planktivorous and piscivorous seabird colonies in Spitsbergen. Polar Biol 36:363–372

Chapter 9
Cryosols of Antarctica

9.1 Introduction

Only 0.35 % of Antarctica is ice-free, amounting to an area of 49,500 km^2 (Fig. 9.1). The ice-free areas are distributed from greatest to least: the Transantarctic and Pensacola Mountains (53 %), the Antarctic Peninsula (20 %), MacRobertson Land (11 %), and Queen Maud Land (7 %), with the remaining areas comprising 9 % (Table 9.1).

Antarctica is divided by the Transantarctic Mountains into what is commonly known as East Antarctica, which contains the massive East Antarctic ice sheet (EAIS) over bedrock, and West Antarctica, which contains the marine-based West Antarctic ice sheet (WAIS). These two ice sheets contain 70 % of the Earth's freshwater and have mean elevation of over 3,000 m. While the EAIS generally has been stable during the Pleistocene, the WAIS disintegrated during Northern Hemisphere glaciations (Denton et al. 1991).

9.2 Soil-Forming Factors

9.2.1 Climate, Permafrost, and Active-Layer Thickness

There are three distinct climates in Antarctica. The western Antarctic Peninsula and the offshore islands (South Shetland and South Orkney Islands) have a subantarctic maritime climate with mean annual air temperatures ranging from −2 to −4 °C and mean annual precipitation between 400 and 700 mm (water equivalent) (Table 9.2). The coastal regions of East Antarctica have a polar desert climate with a mean temperature of −10 to −12 °C and a mean annual precipitation of 200–250 mm. The inland mountains have a cold desert climate with a mean temperature of −20 to −35 °C and a mean annual precipitation of <50 mm.

J.G. Bockheim, *Cryopedology*, Progress in Soil Science,
DOI 10.1007/978-3-319-08485-5_9

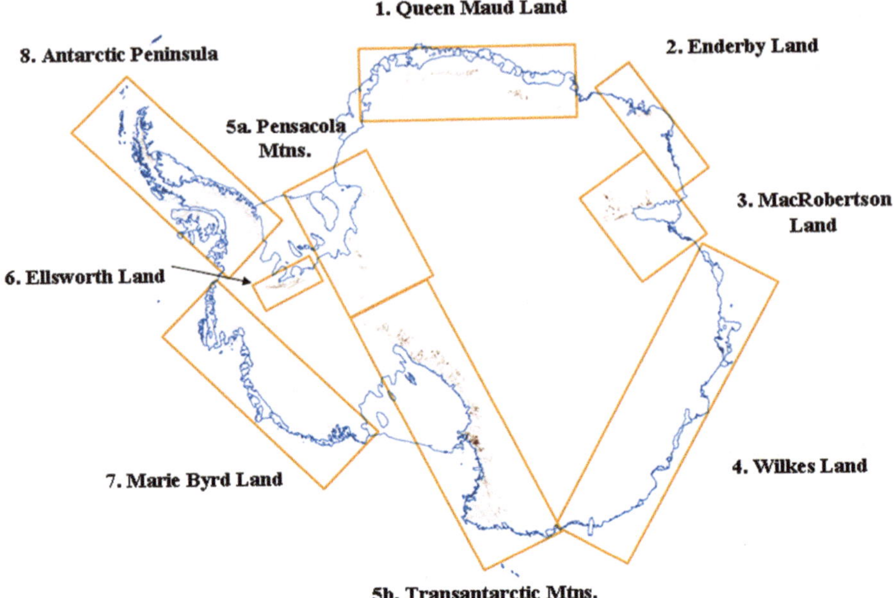

Fig. 9.1 Ice-free regions of Antarctica

Table 9.1 Ice-free regions of Antarctica

No.	Subregion	Key ice-free areas	Area (km^2)
1	Drønning Maud Land	Maudeheimvidda, Fimbulheimen, Thorshavnheiane	3,100
2	Enderby Land	Scott-Tula-Napier Mtns.	830
3	MacRobertson Land	Prince Charles Mountains, Mawson Escarpment	3,500
4	Wilkes Land	Bunger Hills, Windmill Is.	490
5a	Pensacola Mtns.	Shackleton Range, Pensacola Mtns., Thiel Mtns., Pagano Nunatak	1,330
5b	Transantarctic Mtns.	N. Victoria Land, McMurdo Dry Valleys, S. Victoria Land	20,000
6	Ellsworth Mtns.	Sentinel Range, Heritage Range	1,150
7	Queen Maud Land	Executive Committee Range, Crary Mtns.	1,000
8	Antarctic Peninsula	S. Shetland Is., S. Orkney Is., Alexander Is., Antarctic Peninsula	13,600

Permafrost is continuous in continental Antarctica and discontinuous in the South Shetland Islands (Fig. 2.6). From limited data, the permafrost thickness in Antarctica ranges from less than 100 m in the South Shetland Islands to more than 1,000 m in the MDV (Bockheim 1995). The temperature at the top of the permafrost (TTOP) ranges between 0 and −2.5 °C in the South Orkney and South Shetland Islands and on the western Antarctic Peninsula (WAP) mainland from the northern

Table 9.2 Climate data for selected stations in ice-free subregions of Antarctica

Sub-region	Station	Latitude (S)	Longitude	Elev. (m)	MAAT (°C)	Mean temp. Dec., Jan. (°C)	MAP (mm)
1	Neumayer	70.683°	8.266W	40	−17.0	−4.7	~400
	Sanae	71.687°	2.842°W	805	−17.1	−4.7	~100
2	Syowa	69.000°	39.583°E	15	−10.5	−1.0	
	Molodezhnaya	66.275°	100.160°E	42	−11.0	−1.0	250
	Mawson	67.000°	62.883°E	8	−11.2	−0.1	~200
3	Davis	68.583°	77.967°E	12	−10.3	−0.3	~200
	Mawson	67.600°	62.867°E	10	−11.2	0.0	
	Zhong Shan	69.367°	76.367°	15	−9.2	0.8	~200
	Grove Mtns.	73.25°	74.55°E	2,160		−18.5	
4	Casey	66.279°	110.536°E	12	−9.2	−0.4	223
	Mirny	66.55°	93.01°E	30	−11.4	−2.0	379
5a	Halley Bay	75.500°	26.650°W	42	−18.7	−5.3	~150
5b	Lake Bonney	77.733°	162.166°W	150	−17.9	nd	<100
	McMurdo	77.880°	166.730°E	24	−17.4	−3.6	202
	Lake Vanda	77.517°	161.677°E	85	−19.8	0.8	5
6	Ellsworth	77.700°	41.000°W	42	−22.9	−8.0	~150
	Sky-Blu	74.79°	71.48°W	1,510	−19.8	nd	nd
8	Signy Island	60.700°	45.593°W	90	−3.4	0.9	400
	Orcadas	60.750°	44.717°W	12	−2.8	−0.1	486
	King George I.	62.233°	58.667°W	12	−2.0	1.1	635
	Livingston I.	62.650°	60.350°W	35	−1.7	nd	800
	Esperanza	63.4°	56.98°W	13	−5.0		423
	Palmer	64.767°	64.005°W	8	−2.4	2.0	679
	Rothera	67.567°	68.013°W	33	−3.4	0.8	768
	Fossil Bluff	71.333°	68.283°W	55	−8.6	nd	nd
	Marambio	64.234°	56.625°W	5	−8.9	−1.7	250

Source: Schwerdtfeger (1984); various station climate summaries

tip to at least 67°S (Table 9.3). Permafrost temperatures range from −2.6 to −10 °C in areas on the southern WAP to Alexander Island, in the Weddell Sea islands, and in coastal East Antarctica (regions 1, 2, 3, 4, and 7). Permafrost temperatures range from −13 to −23 °C in the Thiel Mountains and Pensacola Mountains (region 5a), Transantarctic Mountains (TAM; region 5b), the Queen Maud Land mountains (region 1), and possibly in the southern Prince Charles Mountains and Grove Mountains (region 3). Dry permafrost occurs primarily in these mountains, especially those in central and southern Victoria Land.

Active-layer (seasonal thaw layer) depths are dependent on the regional climate. In the South Orkney and South Shetland Islands, the active-layer depth commonly ranges between 1.0 and 2.0 m (Table 9.3). The active-layer depth in East Antarctica

Table 9.3 Active-layer depths and permafrost temperatures for selected stations in ice-free subregions

Sub-region	Station	Latitude (°S)	Longitude (°)	Elev. (m)	Active-layer depth (m)	Permafrost temp. (°C)[a]
1	Troll	72.011	2.533E	1,335	0.08	−17.8
	Sanae	71.687	2.842°W	805	0.15	−16.8
	Novozalarevskaya	70.763	11.795E	80	0.7	−9.7
	Aboa	73.033	13.433°W	450	0.6	nd
	Farjuven Bluffs	72.012	3.388°W	1,220	0.25	−17.8
	Sør Rondane Mtns.	71.500	24.5E	1,250	0.1–0.4	nd
2	Syowa	69.000	39.583°E	15	nd	−8.2 (6.8)
	Molodezhnaya	66.275	100.760°E	7	0.9–1.2	−9.8
3	Progress	69.404	76.343	96	>0.5	−12.1
	Grove Mtns.	79.920	74E	1,200	0.2	nd
	Larsemann Hills	69.400	76.27E	50	1.0–1.1	nd
4	Casey Station	66.280	110.52E	10–100	0.3–0.8	nd
5a						
5b	Simpson Crags	74.567	162.758E	830	0.35	nd
	Oasis	74.700	164.100E	80	1.6	−13.5
	Mt. Keinath	74.558	164.003E	1,100	nd	nd
	Boulder clay	74.746	164.021E	205	0.25	−16.9
	Granite Harbour	77.000	162.517E	5	0.9	nd
	Marble Point	77.407	163.681E	85	0.4	−17.4
	Victoria Valley	77.331	161.601E	399	0.24	−22.5
	Bull Pass	77.517	161.850E	150	0.5	−17.3
	Minna Bluff	78.512	166.766E	35	0.23	−17.4
	Scott Base	77.849	166.759E	80	0.30	−17.0
6	Ellsworth Mtns.	78.500	85.6W	800–1,300	0.15–0.50	nd
7	Russkaya	74.763	136.796	76	0.1	−10.4
8	Signy Island	60.700	45.583W	90	0.4–2.2	−2.4
	King George Island	62.088	58.405W	37	1.0–2.0	−0.3 to −1.2
	Deception-Livingston Is.	62.670	60.382W	272	1.0	−1.4 to −1.8
	Cierva Point	64.150	69.95W	182	2.0–6.0	−0.9
	Amsler Island, Palmer	64.770	64.067W	67	14.0	−0.2
	Rothera	67.570	68.13W	32	1.2	−3.1
	Marambio Station	64.240	56.67W	5–200	0.6	nd

Source: Vieira et al. (2010); various reports
[a]Depth of measurement in parentheses where available

ranges between 0.5 and 1.0 m. In the Transantarctic Mountains, active layer depths range between 0.3 and 1.0 m, depending on elevation and proximity to the McMurdo coast. Active patterned ground is present throughout ice-free areas of Antarctica.

9.2.2 Biota

Vegetation is an important soil-forming factor in Antarctica by virtue of its presence or its absence. Two higher plants (*Deschampsia Antarctica* and *Colobanthus quitensis* grasses) are found only along the WAP and the South Orkney and South Shetland Islands. Small (10–1,000 m^2) patches of continuous vegetation cover, primarily mosses and lichens, may occur in coastal areas of regions 1, 2, 3, 4, 7, and 8. There are 427 species of lichens in Antarctica, 40 % of which are endemic (Nayaka and Upreti 2005). Algae influence soil development in these same areas. Endolithic lichens produce organic matter and initiate chemical and physical weathering throughout Antarctica. Microorganisms are present in nearly all soils of Antarctica, except possibly in old soils with pronounced salt accumulation.

Birds, primarily penguins but also skua gulls and petrels, contribute organic matter, phosphates, and Na and are important in coastal areas of Antarctica (Beyer 2000; Beyer et al. 2000). Ornithogenic soils are best expressed directly under active Adélie (*Pygoscelis adeliae*), chinstrap (*P. antarctica*), or gentoo (*P. papua*) penguin rookeries but are also commonly found at abandoned rookeries, where ornithogenic soils remain hundreds to thousands of years later (Myrcha and Tatur 1991). About 200 million kg of C and 20 million kg of P are deposited annually in rookeries of maritime Antarctica from Adélie and Chinstrap penguin excrement (Pietr et al. 1983; Myrcha and Tatur 1991). The high levels of seabird manure are a function of nutrient upwelling at the Antarctic Convergence. Along the continental shelf of the Antarctic Peninsula; nutrients feed large blooms of phytoplankton to sustain Antarctic krill, which are subsequently consumed and excreted by seabirds to develop the soils of maritime Antarctica.

9.2.3 Parent Materials and Time

Most of the soils in coastal regions of Antarctica are of Late Glacial Maximum age, or younger. However, strongly developed soils of early Pleistocene to Miocene age occur in the Sør Rondane Mountains (region 1), the southern Prince Charles Mountains and Grove Mountains (region 3), the Thiel Mountains and Pensacola Mountains (region 5a), and the Transantarctic Mountains (region 5b). Parent material results in unique soils in areas with sulfide rocks (sulfuric subgroups of soils), carbonates (calcification in coastal areas and north Victoria Land), and sandy materials (Gelipsamments, Psammorthels, and Psammoturbels).

9.3 Soil Properties

In this section soil properties will be considered for each of the three broad climatic regions, including the western Antarctic Peninsula and the offshore islands, which comprise 1,700 km^2, or 3.4 % of the total ice-free area of the continent; coastal East

Antarctica which as an ice-free area of 1,635 km^2 (3.3 %); and the inland mountains which account for 93.3 % (46,165 km^2).

9.3.1 Western Antarctic Peninsula

Soils at elevations below 30 m in the South Shetland and South Orkney Islands and at elevations as high as 200 m along the western Antarctic Peninsula mainland lack permafrost, are classified as Gelents, Gelepts, and Gelists (Gelists are not formally recognized in ST yet), and are not considered further here. However, most of the soils of the WAP contain permafrost in the upper 1 m, are intensely cryoturbated, and are classified as Turbels. Orthels occur on well-drained uplands, and Histels occur in wet depressions (Simas et al. 2014).

Soils along the WAP contain greater amounts of silt and clay than soils elsewhere in Antarctica, reflecting the greater role of liquid water and greater weathering rates (Table 9.4). Because of their proximity to the coast, soils of the WAP are dominated by Na in water-extracts and on exchange sites. Because of greater plant cover, the SOC concentrations are greater than in soils elsewhere in Antarctica. The abundance of clay and SOC enables soils of the WAP to have moderately high cation exchange capacities. However, leaching from abundant precipitation, including rainfall, results in strongly acid to very strongly acid pH values and low base saturation. Some parent materials contain sulfides that upon weathering yield acid sulfate conditions; these soils are common on the Keller Peninsula of King George Island and on Seymour Island in the Weddell Sea. The ultra-acid pH values in the last pedon of Table 9.5 reflect the sulfurization process (De Souza et al. 2012).

Soils on the WAP often contain very high levels of Mehlich-1 extractable P because of the influence of penguins and migratory birds such as petrels and skua gulls. This process is referred to as phosphatization (Pereira et al. 2013a, b). Birds also produce high levels of N from their guano and soluble salts from their nasal excretions. Some examples of soils occurring to a limited extent along the WAP are given in Fig. 9.2a, d, e. The ornithogenic soil shown in Fig. 9.2f is similar to those along the WAP.

9.3.2 Coastal East Antarctica

Soils of coastal East Antarctica from Syowa and Molodezhnaya stations in Marie Byrd Land to Casey Station in Wilkes Land bear properties intermediate between those of the WAP and Antarctica's inland mountains. These soils are derived from glacial and marine sediments and tend to be shallow and coarse textured (Table 9.6). However, soils of East Antarctica are often influenced by birds (Beyer 2000; Beyer et al. 2000), and thin layers of peat may develop in bedrock depressions. One of the

Table 9.4 Analytical properties of selected taxa from the South Shetland Islands

Horizon	Depth (cm)	Sand (%)	Silt (%)	Clay (%)	Ex. Ca	Ex. Mg	Ex. Na	Ex. K	Ex. Al	CEC	BS (%)	TSS (%)	pH (H_2O)	TOC (%)	Total N (%)	Extr. P (mg/L)
					(cmol$_c$/kg)											
		Typic Aquorthels; Elephant Island; O'Brien et al. (1979)														
C1	0–5	51	39	10	5.4	0.39	0.22	0.06		6.07		0.019	8.2	0.33	0.04	22
C2	5–25	56	44	<1	22.5	0.47	0.22	0.08		23.27		0.023	8.4	0.5	0.03	14
C3	45–55	52	47	<1	40.4	0.51	0.14	0.09		41.14		0.023	8.5	0.61	0.05	20
Cg	65–75	52	48	<1	39.8	0.55	0.14	0.1		40.59		0.024	8.6	0.55	0.05	19
Cf	90															
							(cmol$_c$/dm^{-3})									
		(Ornithic) Glacic Haploturbels; King George Island; Michel et al. (2006)														
	0–10	42	38	20	15.4	5	396	228	0.2	22.7	23		5.9		0.06	33.7
	10–20	39	35	26	7.8	5	388	244	0	15.1	15		7.1		0.01	37.7
	20–30	42	34	24	8.2	2.7	376	224	0	13.1	13		7.8		0	38.3
	30–40	41	35	24	7.8	4.6	320	216	0	14.3	14		8.2		0	38.5
	40–50	45	33	22	7.9	4.4	380	244	0	14.6	15		8		0	40.6
	50–60	40	39	21	7.9	4.4	344	232	0	14.4	14		8.5		0	38.4
		(Ornithic) Lithic Umbriturbels; King George Island; Michel et al. (2006)												dag/kg		
	0–10	70	21	9	3.6	2.6	232	200	5.6	7.7	8		4.8	4.5	0.08	26.6
	10–20	65	26	9	2.2	1.2	216	260	6.6	5	5		4.5	2.8	0.08	30.9
	20–30	67	24	9	1.8	0.72	204	280	5.4	4.1	4		4.3	2.7	0.08	35.9
	30–40	64	26	10	1.8	0.74	152	236	5.2	3.8	4		4.2	2.5	0.1	38
	40–50	63	24	13	1.4	0.6	90	164	3.8	2.8	3		4.2	6	0.12	37.7
	50–60	63	24	13	1.7	0.71	100	170	4.2	3.2	3		4.1	7.4	0.09	34

(continued)

Table 9.4 (continued)

Horizon	Depth (cm)	Sand (%)	Silt (%)	Clay (%)	Ex. Ca (cmol$_c$/kg)	Ex. Mg	Ex. Na	Ex. K	Ex. Al	CEC	BS (%)	TSS (%)	pH (H_2O)	TOC (%)	Total N (%)	Extr. P (mg/L)
		(Ornithic) Lithic Fibristels; King George Island; Michel et al. (2006)														
	0–10	53	27	20	3.3	1.4	184	144	1.8	5.9	6		5	8.6	0.17	32.8
	10–20	57	27	16	3.7	1.3	162	196	3.8	6.2	6		4.8	5.8	0.1	39.8
	20–30	64	24	12	3.8	1.5	204	198	4.6	6.8	7		4.6	4.7	0.15	36.2
	30–40	52	31	17	3.3	1.1	146	182	3.8	5.5	6		4.6	4.7	0.17	28
	40–50	51	34	15	3.7	1.2	174	196	6.2	6.2	6		4.3	7.7	0.13	23.5
		(Ornithic) Psammentic Aquiturbels; King George Island; Michel et al. (2006)														
	10–20	75	14	11	1.2	0.59	146	150	6.8	2.8	3		4.2	3.5	0.1	34.6
	20–30	72	19	9	1.1	0.55	164	162	6.8	2.8	3		4.2	2.7	0.04	29.9
	30–40	76	14	10	1.3	0.62	158	150	8.8	3	3		4.1	1.2	0.04	38.9
	40–50	76	15	9	1.4	0.65	128	168	9.8	3	3		4.1	1	0.1	35.1
		Typic Haploturbels; King George Island; Simas et al. (2007)														(mg/dm^3)
C1	0–10	18	45	37	22.5	15.3	1,360	115	0	48	92		6.6	0.1		65
C2	10–20	22	52	25	29.8	21.4	1,440	148	0	60	96		7	0.1		48
		Typic Psammoturbels; King George Island; Simas et al. (2007)														
O	0–10	24	46	30										2.2	0.35	
A	10–20	29	44	27	2	0.3	332	680	7.5	40	14		4.2	1	0.23	1,325
Bw	20–30	36	47	17	3.1	0.7	286	660	9.8	45	15		4.1	0.5		1,143
BC1	30–40	34	44	22	4.9	1.5	274	680	11.3	50	19		4.1	0.5		849
BC2	40–50	36	46	18	5.8	2.5	244	660	13.5	47	28		4.2	0.4	0.08	742

		Andic Umbriturbels; King George Island; Simas et al. (2007)														
A1	0–10	30	38	30	8.5	5.6	290	132	2.7	33	48		4.6	2.4	0.4	2,466
A2	10–20	27	41	27	4.7	2	306	184	2.6	25	34		4.6	2	0.42	3,910
Bw	20–30	29	37	17	7.7	1.8	320	380	3.1	40	30		4.5	1.3	0.35	4,156
BC	30–40	26	42	22	8.9	1.3	310	410	4.5	41	32		4.4	1.3	0.44	3,242
C	40–60	20	40	18	9.6	1.2	280	300	8.4		31		4.1	1.2		1,932
		Typic Fibristels; King George Island; Simas et al. (2007)														
Oi1	0–10	17	49	34	2.8	0.8	150	80	7.2	27	17		4.3	4.4	0.64	658
Oi2	10–20	23	51	26	1.8	0.4	151	72	7.6	30	10		4.2	3.5	0.49	667
Oi3	20–30	32	39	29	2.2	0.5	170	101	12.6	36	10		4.2	2.7	0.32	904
Oi4	30–40	29	39	32	2.5	0.6	155	81	12.1	40	10		4.3	2.9	0.22	1,117
Oi5	40–50	23	42	35	2.6	0.6	172	81	11.7	36	12		4.3	3.3	0.3	1,030
												EC				
		Sulfuric Aquorthels; Seymour Island; Souza et al. (2014)										(dS/m)				
A	0–5	54	24	21	4.7	10.1	24.2	5.1	0.29	31	28	4.5	5.1	0.33		24.2
AC	5–35	58	39	2	25.1	9.9	27.9	7.1	0	44	44	4.1	7.1	0.65		27.9
C	35–40	43	25	31	2.5	12.1	53.8	3.3	6.2	39	23	3.4	3.3	1.3		53.8
Cg1	40–55	18	23	58	5.1	15.4	57.7	3.1	12.6	60	33	4.8	3.1	0.65		57.7
Cg2	55–88	51	44	4	19.6	10.7	21.5	3.6	6.7	55	41	4.9	3.6	0.65		21.5
Average mineral		**46**	**34**	**19**	**9.6**	**4.1**	**261.7**	**213.9**	**4.8**	**25.76**	**23**	**2.4**	**5.5**	**1.7**	**0.12**	**575**

Table 9.5 Analytical properties of selected taxa from East Antarctica

Horizon	Depth (cm)	Sand (%)	Silt (%)	Clay (%)	BS (%)	EC (dS/m)	pH ($CaCl_2$)	TOC (%)	Total N (%)	Extr. P (mg/L)
		Typic Spodorthels; Casey Station; Blume and Bölter (2014)								
O	0–6									
AE	6–8	88	7	5	65	3.8	4.2	3.9	0.16	0.33
Bh	8–11	83	12	5	60	1.5	4.3	4.1	0.37	0.47
Bsh1	11–18	67	19	13	67	1.1	4.4	2.4	0.48	0.86
Bsh2	18–30	67	19	13	67	1.1	4.6	2	0.36	
Bsh3	30–40	78	17	5	76	2	4.8	9	0.41	1.36
Cg	40–50				93	1.4	5.3	3	0.03	
		Lithic Folistels; Casey Station; Blume and Bölter (2014)								
Oe	0–3				88	3.5	4.7	29	1.5	0.18
Oa1	3–8				73	2.8	3.9	20	2.2	0.17
Oa2	8–16				60	0.8	3.8	19	2.1	0.25
Oa/C	16–28				46	0.4	3.8	16	1.3	0.23
R	28									
		Lithic Fibristels; Casey Station; Blume and Bölter (2014)								
Oi1	0–1				85	1.8	6	27	1.4	2.1
Oi2	1–8				68	4.9	5.3	26	1.2	0.63
Oi3	8–18				83	0.6	4.9	16	0.89	0.01
R	18									
	(Ornithic) Lithic Haplorthels; Larsemann Hills; Mergelov et al. (2014)									
							(H_2O)			
O/A	0–1					0.27	6.1	8.4	0.5	
A	1–4					0.43	6	15.5	0.9	
Bw	4–15					0.05	6.9	0.7	0.1	
R	15									
		Typic Aquorthels; Larsemann Hills; Mergelov et al. (2014)								
							(H_2O)			
	0–1					0.12	8.4	0.37	0.041	
	1–3					0.074	7.2	0.28	0.035	
	3–15					0.04	7.5	0.12	0.019	

unusual discoveries in Antarctic pedology was the identification of Spodorthels on abandoned penguin rookeries near Casey Station (Beyer and Bölter 2000). Whereas the Spodorthels and Histels of East Antarctica are very strongly to extremely acid, soils lacking the influence of birds or buildup of organic materials are neutral to weakly alkaline. Soils of coastal East Antarctica have levels of silt, clay, SOC, total N, CEC, and soluble salts that are intermediate between those of the WAP and the inland mountains. Examples of soils occurring to a limited extent in coastal East Antarctica are given in Fig. 9.2b, c.

Fig. 9.2 Soils of limited extent along the western Antarctic Peninsula and in coastal East Antarctica: (**a**) Sulfuric Haplorthel on Seymour Island (photo by C. Schaefer); (**b**) Typic Spodorthel in abandoned penguin rookery, Casey Station, Wilkes Land (photo by H.-P. Blume and M. Bölter); (**c**) Typic Aquorthel in the Larsemann Hills (photo by N. Mergelov); (**d**) Fibrist at Cierva Point, western Antarctic Peninsula (photo by J. Bockheim); (**e**) Saprist at Cierva Point (photo by J. Bockheim); (**f**) ornithogenic soil, Cape Hallett, north Victoria Land (photo by M. Balks)

9.3.3 Inland Mountains

The inland mountains by far account for the most ice-free area of Antarctica (93 %). Although there are longitudinal climatic gradients in valleys extending from the Transantarctic Mountains to the Ross Sea, the properties of soils in the inland mountains generally reflect hyperarid and hypergelic conditions. The soils of the inland mountains generally have low amounts of silt and clay, very low quantities of SOC, abundant soluble salts, and alkaline pH values (Table 9.6). Examples of common and less common soils in the inland mountains are shown in Figs. 9.3 and 9.4, respectively.

9.4 Soil-Forming Processes

These soil-forming factors lead to soil-forming processes that vary in magnitude by ice-free region in Antarctica. The relative importance of each of these processes is indicated by one asterisk (*) as being unimportant or absent, two asterisks (**) as

Table 9.6 Chemical and physical properties of selected soil taxa from the Transantarctic Mountains

Horizon	Depth (cm)	EC (dS/m)	mmol$_c$/L								Ex. Na (%)	Fe$_d$ (%)	>2 mm (%)	Sand (%)	Silt (%)	Clay (%)
			pH	Na^+	Ca^{2+}	Mg^{2+}	K^+	Cl^-	SO_4^{2-}	NO_3^-						
			75-06 Typic Anhyorthels													
Bw1	1–12	4.7	6.6	508.9	0.94	0.93	0.11	4.34	488.0	18.6	99.6	0.18	55	90	2.7	7.3
Bw2	12–18	1.56	6.6	163.1	0.49	0.43	0.05	1.37	156.0	6.7	99.4	0.16	45	96.9	2.3	0.8
Bw3	18–54	1.2	5.3	124.0	0.42	0.30	0.06	1.04	117.8	5.9	99.4	0.17	85	94.4	3.3	2.3
Cn	54–115	0.33	6.4	32.6	0.26	0.17	0.01	0.24	32.2	0.6	98.6	0.14	25	96.3	2.3	1.4
			75-14 Typic Haploturbels													
D	0–1	0.84	8.3	6.0	0.94	0.35		5.5	2.1	0.0	82.3		nd			
Cn1	1–5	0.88	9.2	7.0	0.11	0.19		5.1	1.7	0.5	95.9		nd			
Cn2	5–25	0.2	7.5	1.5	0.02	0.05		1.2	0.1	0.2	95.3		nd			
Cf	25+	0.135	6.9	0.9	0.13	0.11		0.6	0.3	0.2	79.2		nd			
			76-43 Lithic Anhyorthels													
D	0–1	1.05	5.9	25.7	8.5	4.9	0.8	0.8	11.7	27.28	64.4		nd	93	4	3
Bw	1–7	5	7.4	116.4	47.2	40.3	2.4	4.7	79.4	122.1	56.4		nd	79	16	5
Bwz	7–15	26	7.8	967.9	13.5	130.8	5.8	8.9	78.1	1,031	86.6		nd	72	17	11
Cn1	15–33	1.25	8.4	30.9	3.5	15.6	2.2	6.3	3.4	42.49	59.1		nd	81	16	3
Cn2	33–52	0.75	8.8	15.2	3.9	9.9	1.0	4.2	3.6	22.14	50.7		nd	85	13	2
			76-52 Typic Haplorthels													
D	tr	2.3	8.4	58.7	7.2	16.6	3.1	65.2	9.2	11.38	68.5		80	92	5	3
Bw	0–12	4.75	8.5	147.9	5.5	44.4	7.2	154.0	22.9	28.11	72.1		55	85	10	5
Bwy	12–20	11	8.8	584.0	8.9	65.8	7.5	28.2	208.3	429.8	87.7		58	84	9	7
Cn1	20–47	1.45	8.7	39.1	3.5	11.5	4.3	26.1	1.4	30.93	67.0		33	86	12	2
Cn2	47+	0.8	8.7	21.3	3.1	5.8	2.3	1.5	1.6	29.46	65.5		nd	89	8	3

			77-36 Glacic Haploturbels													
Cn	0–8	0.62	8.2	2.8	2.3	1.3	0.2	4.4	0.8	0.49	42.4	0.124	nd	97.6	0.8	1.6
Cf	8–16	0.6	7.7	2.5	2.3	1.4	0.2	4.3	0.8	0.48	39.1	0.196	78	97.2	0.3	2.5
Wf	16+	0.069	7.4	0.2	0.5	0.1	0.0	0.2	0.9	0.006	25.9		0			
			79-10 Typic Anhyturbels													
Bw	0.5–10	2.4	6.7	10.0	18.0	2.1	0.2	5.4	20.0	2.3	33.0		58			
Cn1	10–64	5.2	6.0	42.0	7.9	5.8	0.4	36.0	8.5	9.5	74.9		75			
Cn2	64–100	2.2	6.5	19.0	0.8	1.8	0.2	17.0	2.3	2.3	87.4		49			
													43			
			83-41 Petrosalic Anhyorthels													
Cn	0–14	5.3	7.3	47.3	13.0	1.6	0.6	32.4	18.8	1.67	75.8		65			
2Bwzb	14–28	80	7.8	1,088.8	36.9	110.9	2.6	747.6	227.3	77.03	87.9		53			
2Bwb	28–70	9.3	7.1	100.4	14.2	10.1	1.0	78.1	28.3	1.79	79.8		43			
3Bwb	70–120	3.6	6.4	16.2	9.3	13.5	1.6	17.0	17.3	1.59	39.9		35			
			84-47 Petronitric Anhyorthels													
Bw1	0–4	4.8	7.0	21.9	34.4	7.6	1.1	18.2	45.9	3.64	33.7		14			
Bwz1	4–20	80	7.1	820.4	16.1	83.7	7.5	947.8	157.6	23.63	88.4		40			
Bwz2	20–42	31	6.9	302.2	6.4	55.0	6.6	269.2	50.4	20.05	81.6		48			
Bw2	42–85	4.4	6.7	30.3	2.3	7.4	2.7	30.6	7.7	4.75	71.1		78			
	Avg.	**9.2**	**7.4**	**167.3**	**8.5**	**20.3**	**2.2**	**79.0**	**56.4**	**61.1**	**71.5**	**0.16**	**51**	**89**	**8**	**4**

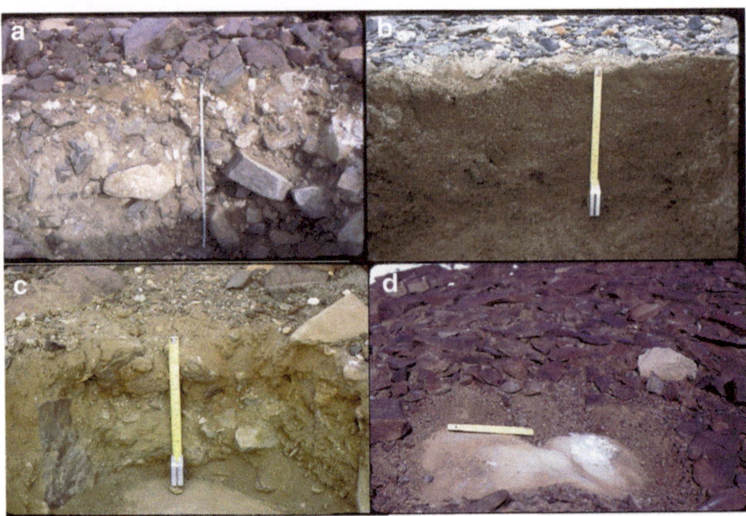

Fig. 9.3 Soils of the inland Antarctic mountains: (**a**) Typic Anhyorthel (pedon 78–11) in the Britannia Range, Darwin Glacier area; (**b**) Typic Haploturbel (pedon 75–14) in eastern Taylor Valley of the McMurdo Dry Valleys; (**c**) Typic Anhyturbel (pedon 78–29) in the Darwin Mountains; (**d**) Glacic Haploturbel (pedon 81–27) in the Lichen Hills, north Victorial Land (photos **a**, **b** and **c** by J. Bockheim and photo **d** by S. Wilson)

Fig. 9.4 Soils of limited extent in the inland Antarctic mountains: (**a**) Petrosalic Anhyorthel (pedon 83–41) in central Wright Valley in the McMurdo Dry Valleys; (**b**) Petronitric Anhyorthels (pedon 84–47) in central Wright Valley; (**c**) Nitric Anhyorthels (pedon 80–18) on Mount Fleming, McMurdo Dry Valleys; (**d**) Lithic Anhyturbels (pedon 80–09) in Arena Valley, McMurdo Dry Valleys. Photos **a** and **b** by S. Wilson and **c** and **d** by J. Bockheim

being of moderate importance, and three asterisks (***) as being of major importance in Antarctica (Table 9.7). The table lists each of the ice-free regions and subdivides them, where appropriate, into coastal and inland (mountain) sub-regions. Processes such as rubification, salinization, desert pavement formation, and permafrost development operate to the greatest extent in the inland mountains of regions 1, 3, 5a, 5b, 6 and possibly 7. Calcification is not a dominant process and occurs primarily in coastal areas of regions 2, 3, 5, and 7. Soil organic matter accumulation, acidification, hydromorphism, phosphatization, paludification, and pervection occur in coastal areas of regions 1 through 4, 7, and 8 but also in the mountains of the Antarctic Peninsula. In Antarctica, as in the Arctic, hydromorphism leads to reductive Eh values but no apparent redoximorphic features or gleying. Sulfurization is restricted to areas with sulfide-enriched parent materials, such as King George Island and Seymour Island. Podzolization is restricted to abandoned penguin rookeries in coastal areas of regions 4 and 8 but may occur to a limited extent in other ice-free areas.

Figure 9.5 shows a gradient in soil-forming processes in Antarctica. Whereas processes such as acidification, clay formation, brunification (melanization), organic matter accumulation, redoximorphism, and podzolization decrease from the subantarctic tundra through the subpolar desert to the cold desert inland mountains, processes such as salinization, alkalization, desert pavement formation, and permafrost thickness increase along the gradient.

9.5 Soil Classification and Geography

Typic Anhyorthels are the dominant soil subgroup comprising nearly 15,000 km^2, or 30 % of the soils in Antarctica (Table 9.8). These soils occur primarily in central and southern Victoria Land (region 5b), but also in the Thiel and Pensacola Mountains and Shackleton Range (region 5a), the Prince Charles Mountains (region 3) and the mountains of Queen Maud Land (region 1). Typic Haploturbels and Typic Anhyturbels occupy 14 and 13 % of the soils of ice-free regions of Antarctica, respectively. Most abundant in central Victoria Land, they are common in most mountainous regions of Antarctica. Soils in lithic subgroups comprised only 15 % of the soils; however, in the mountains of Antarctica, especially regions 1, 3, 5a, 5b, 6, 7, and 8, we were unable to differentiate the Rockland land type from soils in lithic subgroups so that we have probably underestimated the areal distribution of lithic soils. Typic Gelorthents occupy about 8 % of the ice-free areas of Antarctica, mainly in Palmer and Graham Lands but also in the SSI and SOI (region 8).

Forty-four percent of the soils of Antarctica are Orthels, gelisols that show minimal evidence of cryoturbation and occur in dry landscape positions; 36 % of the soils are Turbels showing cryoturbation in more moist landscape positions (Table 9.9). Only 16 % of the soils of Antarctica lack permafrost in the control section and are classified as Entisols (Gelorthents), Inceptisols (Haplogelepts,

Table 9.7 Relative importance of soil-forming processes in ice-free regions and sub-regions of Antarctica[a]

Soil-forming process	Ice-free region[b]													
	1C	1M	2C	3C	3M	4C	5aM	5bC	5bM	6M	7C	7M	8C	8M
Rubification	*	***	*	*	***	**	***	*	***	*	*		*	**
Salinization	*	***	*	*	***	*	***	*	***	*	*		*	*
Calcification	*	*	**	**	*	*	*	**	*	**	**		*	*
Soil organic matter accumulation	**	*	**	**	*	**	*	*	*	*	*		**	**
Pervection	**	*	**	**	*	**	*	*	*	*	*		**	**
Desert pavement formation	**	***	*	**	***	*	***	**	***	**	**		*	*
Permafrost development	**	***	**	**	***	**	***	**	***	***	***		*	**
Acidification	*	*	**	*	*	***	*	*	*	*	**		***	**
Hydromorphism	**	*	***	***	*	***	*	*	*	*	**		***	**
Phosphatization	*	*	**	**	*	**	*	*	*	*	*		***	***
Sulfurization	**	*	**	*	*	**	*	*	*	*	*		**	*
Paludification	*	*	**	**	*	**	*	*	*	*	*		***	**
Podzolization	*	*	*	*	*	**	*	*	*	*	*		**	**

[a]Relative importance: *low or absent; **moderate; ***important
[b]*C* coastal, *M* inland mountains

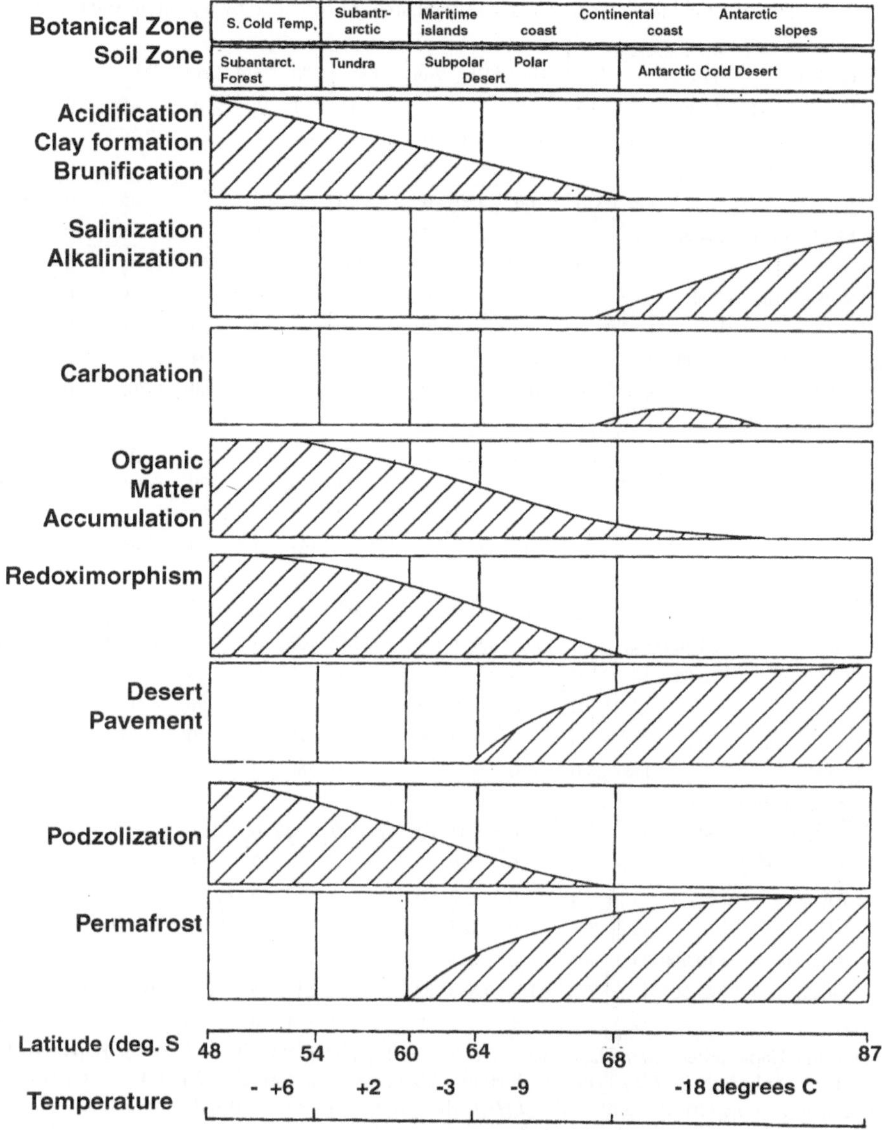

Fig. 9.5 Gradient in soil-forming processes in Antarctica (Goryachkin et al. 2004)

Humigelepts, Dystrogelepts), or Histosols (Cryofibrists, Cryohemists, Cryosaprists, and Cryofolists). These soils occur almost exclusively along the western Antarctic Peninsula and at elevations below 50 m in the SSI and SOI; these organic soils may contain permafrost below 2 m. We suggest that ornithogenic soils occupy only 0.5 % of ice-free areas in Antarctica, but this may be an underestimate.

Table 9.8 Distribution (area and percentage of total area) of soil taxa[a] by region in Antarctica

	Region	Approx. area (km²)		LHt	GHt	THt	AqHt	TAt	LAt	LAqt	LAo	TAo	THo	LHo
1	Queen Maud Land	**3,400**	%	0	5	24	0	15	2	0	4	45	0	0
			km²	0	170	816	0	510	68	0	136	1,530	0	0
2	Enderby Land	**1,500**	%	25	4	30	0	0	0	3	0	3	0	30
			km²	375	60	450	0	0	0	45	0	45	0	450
3	MacRobertson Land	**5,400**	%	0	17	6	0	23	0	0	7	36	0	0
			km²	0	918	324	0	1,242	0	0	378	1,944	0	0
4	Wilkes Land	**700**	%	0	0	7	0	0	0	7	0	0	0	58
			km²	0	0	49	0	0	0	49	0	0	0	406
5a	Pensacola Mtns.	**1,500**	%	0	7	9	0	17	0	0	4	47	0	0
			km²	0	105	135	0	255	0	0	60	705	0	0
5b	Transantarctic Mtns.													
	NVL	2,420	%	0	36	0	0	36	0	0	14	14	0	0
			km²	0	871	0	0	871	0	0	339	339	0	0
	CVL	10,890	%	0	2	36	0	14	2	0	1	43	0	0
			km²	0	218	3,920	0	1,525	163	0	54	4,683	0	0
	SVL	10,890	%	0	7	9	0	17	0	0	4	47	0	0
			km²	0	762	980	0	1,851	0	0	436	5,118	0	0
	Subtotal	**24,200**	km²	0	1,851	4,901	0	4,247	163	0	829	10,140	0	0
6	Ellsworth Mtns.	**2,100**	%	0	0	0	0	0	0	0	36	27	14	18
			km²	0	0	0	0	0	0	0	756	567	294	378
7	Marie Byrd Land	**700**	%	0	0	0	10	0	40	0	40	0	0	0
			km²	0	0	0	70	0	280	0	280	0	0	0
8	Antarctic Peninsula													
	S. Orkney, S. Shetland Is.	645	%	4	0	4	2	0	0	0	0	0	41	0
			km²	26	0	26	13	0	0	0	0	0	264	0
	Palmer, Graham Lands	9,355	%	4	0	2	0	0	0	0	0	0	4	6
			km²	374	0	187	0	0	0	0	0	0	374	561
	Subtotal	**10,000**	km²	400	0	213	13	0	0	0	0	0	638	561
	Grand total	**49,500**	km²	775	3,104	6,888	83	6,254	511	94	2,439	14,931	932	1,795
			%	2	6	14	0	13	1	0	5	30	2	4

[a]*LHt* Lithic Haploturbels, *GHt* Glacic Haploturbels, *THt* Typic Haploturbels, *AqHt* Aquic Haploturbels, *TAo* Typic Anhyorthels, *THo* Typic Haplorthels, *LHo* Lithic Haplorthels, *GHo* Glacic Haporthels, Spodorthels, *Orn* Ornithogenic soils, *LHs* Lithic Hemistels, *LFs* Lithic Fibristels, *TGe* Typic Lithic Gelisaprists, *Lcfoh* Lithic Gelifolists

GHo	SAo	NAo	PnAo	TSpo	Orn	LHs	LFs	TGe	THi	LHi	TGqe	LCsh	LCfoh	Other	Total
0	0	0	0	0	0	0	0	0	0	0	0	0	0	5	
0	0	0	0	0	0	0	0	0	0	0	0	0	0	170	3,400
0	0	0	0	0	3	2	0	0	0	0	0	0	0	0	
0	0	0	0	0	45	30	0	0	0	0	0	0	0	0	1,500
0	4	0	0	0	2	0	0	0	0	0	0	0	0	5	
0	216	0	0	0	108	0	0	0	0	0	0	0	0	270	5,400
0	0	0	0	14	7	7	0	0	0	0	0	0	0	0	
0	0	0	0	98	49	49	0	0	0	0	0	0	0	0	700
3	0	3	3	0	0	0	0	0	0	0	0	0	0	7	
45	0	45	45	0	0	0	0	0	0	0	0	0	0	105	1,500
0	0	0	0	0	0	0	0	0	0	0	0	0	0	0	
0	0	0	0	0	0	0	0	0	0	0	0	0	0	0	2,420
0	0	0	0	0	0	0	0	0	0	0	0	0	0	3	100
0	0	0	0	0	0	0	0	0	0	0	0	0	0	327	10,890
3	0	3	3	0	0	0	0	0	0	0	0	0	0	7	
327	0	327	327	0	0	0	0	0	0	0	0	0	0	762	10,890
327	0	327	327	0	0	0	0	0	0	0	0	0	0	1,089	24,200
5	0	0	0	0	0	0	0	0	0	0	0	0	0	0	
105	0	0	0	0	0	0	0	0	0	0	0	0	0	0	2,100
0	0	0	0	0	0	0	10	0	0	0	0	0	0	0	
0	0	0	0	0	0	0	70	0	0	0	0	0	0	0	700
0	0	0	0	1	8	2	2	23	6	0	2	3	2	0	
0	0	0	0	6	52	13	13	148	39	0	13	19	13	0	645
0	0	0	0	0	0	0	0	41	13	10	8	5	5	2	100
0	0	0	0	0	0	0	0	3,836	1,216	936	748	468	468	187	9,355
0	0	0	0	6	52	13	13	3,984	1,255	936	761	487	481	187	10,000
477	216	372	372	104	254	92	83	3,984	1,255	936	761	487	481	1,821	49,500
1	0	1	1	0	1	0	0	8	3	2	2	1	1	4	100

TAt Typic Anhyturbels, *LAt* Lithic Anhyturbels, *LAqt* Lithic Aquiturbels, *LAo* Lithic Anhyorthels, *SAo* Salic Anhyorthels, *NAo* Nitic Anhyorthels, *PnAo* Petronitric Anyorthels, *TSpo* Typic Gelorthents, *THi* Typic Humigelepts, *LHi* Lithic Humigelepts, *TGqe* Typic Gelaquents, *LCsh*

Table 9.9 Influence of permafrost on distribution of soils in ice-free areas of Antarctica

Group	Area (km^2)	Area (%)
With permafrost in upper 1–2 m		
Orthels	21,639	43.7
Turbels	17,708	35.8
Histels	175	0.4
Non-Gelisols	7,903	16.0
Ornithogenic soils	254	0.5
Other	1,821	3.7
Total	49,500	100.0

9.6 Summary

Only 0.35 % of Antarctica is ice-free, amounting to an area of 49,500 km^2. There are three distinct climatic zones in Antarctica. The western Antarctic Peninsula and the offshore islands (South Shetland and South Orkney Islands) have a subantarctic maritime climate with comparatively mild temperatures and abundant precipitation, including rainfall. Coastal East Antarctica has cooler temperatures and less precipitation, all of which falls as snow. The inland mountains feature hyperarid and hypergelic conditions. These climate differences are reflected in active-layer thickness and mean annual ground temperatures, which are greatest in maritime Antarctica and least in the mountains. Birds, especially large penguin colonies, play an important role in soil formation and in the ability of sites to become colonized by vegetation. Whereas soils in maritime and East Antarctica tend to be of Last Glacial Maximum or Holocene in age, soils of inland mountains commonly range from mid-Pleistocene to Miocene in age.

The climatic zonation is reflected in the nature of the soils. The amounts of silt and clay, organic C, total N, extractable P, and soil moisture and soil temperature decline from the West Antarctic Peninsula to East Antarctica and then to the inland mountains. Whereas soils of maritime and East Antarctica are often very strongly acidic and low in base cations and soluble salts, soils of the inland mountains are alkaline, strongly base saturated, and contain abundance salts.

The dominant soil-forming processes in maritime West Antarctica are soil organic matter accumulation, acidification, and hydromorphism, along with sulfurization and phosphatization in some locations. These same processes occur in coastal East Antarctica, along with a weak podzolization process. Desert pavement formation, salinization, and rubification are key processes in the inland mountains.

Because the inland mountains comprise over 93 % of the ice-free area of Antarctic, the predominant soil great groups are Anhyorthels, Haploturbels, and Anhyturbels. Orthels, Turbels, and Histels comprise 44, 36 and 0.4 % of the soils of Antarctica, with nongelisols accounting for 16 %.

References

Beyer L (2000) Properties, formation, and geo-ecological significance of organic soils in the coastal region of East Antarctica (Wilkes Land). Catena 39:79–93

Beyer L, Bölter M (2000) Chemical and biological properties, formation, occurrence and classification of Spodic Cryosols in a terrestrial ecosystem of East Antarctica (Wilkes Land). Catena 39:95–119

Beyer L, Bölter M, Seppelt RD (2000) Nutrient and thermal regime, microbial biomass, and vegetation of Antarctic soils in the Windmill Islands region of East Antarctica (Wilkes Land). Arctic Antarct Alp Res 32:30–39

Blume H-P, Bölter M (2014) Soils of Wilkes Land (the Windmill Islands). In: Bockheim JG (ed) Soils of Antarctica, World soils book series. Springer, New York (in press)

Bockheim JG (1995) Permafrost distribution in the Southern Circumpolar Region and its relation to the environment: a review and recommendations for further research. Permafr Periglac Process 6:27–45

De Souza JJLL, Schaefer CEGR, Abrahão WAP, de Mello JWV, Simas FNB, da Silva J, Francelino MR (2012) Hydrogeochemistry of sulfate-affected landscapes in Keller Peninsula, maritime Antarctica. Geomorphology 155–156:55–61

Denton GH, Prentice ML, Burkle LH (1991) Cenozoic history of the Antarctic ice sheet. In: Tingey RJ (ed) The geology of Antarctica. Clarendon, Oxford, pp 365–433

Goryachkin SV, Blume H-P, Beyer L, Campbell I, Claridge G, Bockheim JG, Karavaeva NA, Targulian V, Tarnocai C (2004) Similarities and differences in Arctic and Antarctic soil zones. In: Kimble JM (ed) Cryosols: permafrost-affected soils. Springer, New York, pp 49–70

Mergelov NS, Konyushkov DE, Lupachev AV, Goryachkin SV (2014) Soils of MacRobertson Land. In: Bockheim JG (ed) Soils of Antarctica, World soils book series. Springer, New York

Michel RFM, Schaefer CEGR, Dias LE, Simas FNB, Benites VM, Mendonça ES (2006) Ornithogenic gelisols (cryosols) from maritime Antarctica: pedogenesis, vegetation, and carbon studies. Soil Sci Soc Am J 70:1370–1376

Myrcha A, Tatur A (1991) Ecological role of the current and abandoned penguin rookeries in the land environment of the maritime Antarctic. Pol Polar Res 12:3–24

Nayaka S, Upreti DK (2005) Schirmacher Oasis, East Antarctica, a lichenologically interesting region. Curr Sci 89:1069–1071

O'Brien RMG, Romans JCC, Robertson L (1979) Three soil profiles from Elephant Island, South Shetland Islands. Br Antarct Surv Bull 47:1–12

Pereira TTC, Schaefer CEGR, Ker JC, Almeida CC, Almeida ICC, Pereira AB (2013a) Genesis, mineralogy and ecological significance of ornithogenic soils from a semi-desert polar landscape at Hope Bay, Antarctic Peninsula. Geoderma 210:98–109

Pereira TTC, Schaefer CEGR, Ker JC, Almeida CC, Almeida ICC (2013b) Micromorphological and microchemical indicators of pedogenesis in ornithogenic cryosols (Gelisols) of Hope Bay, Antarctic Peninsula. Geoderma 194:311–322

Pietr SJ, Tatur A, Myrcha A (1983) Mineralization of penguin excrements in the Admiralty Bay region (King George Island, South Shetland Islands, Antarctica). Pol Polar Res 4:97–112

Schwerdtfeger W (1984) Weather and climate of the Antarctica. Dev Atmos Sci 15:1–262

Simas FNB, Schaefer CEGR, Melo VR, Albuquerque-Filho MO, Michel RFM, Pereira VV, Gomes MRM, da Costa LM (2007) Ornithogenic cryosols from maritime Antarctica: phosphatization as a soil forming process. Geoderma 138:191–203

Simas FNB, Schaefer CEGR, Michel RFM, Francelino MR, Bockheim JG (2014) Soils of the South Orkney and South Shetland Islands, Antarctica. In: Bockheim JG (ed) Soils of Antarctica, World soils book series. Springer, New York (in press)

Souza KKD, Schaefer CEGR, Simas FNB, Spinola DN, de Paula MD (2014) Soil formation in Seymour Island, Weddell Sea, Antarctica. Geomorphology 225:87–99

Vieira G, Bockheim J, Guglielmin M, Balks M, Abramov AA, Boelhouwers J, Cannone N, Ganzert L, Gilichinsky DA, Goryachkin S, López-Martínez J, Meiklejohn I, Raffi R, Ramos M, Schaefer C, Serrano E, Simas F, Sletten R, Wagner D (2010) Thermal state of permafrost and active-layer monitoring in the Antarctic: advances during the International Polar Year 2007–2009. Permafr Periglac Process 21:182–197

References

[illegible]

Chapter 10
Alpine Cryosols

10.1 Introduction

This chapter focuses on cryosols that exist in areas with mountain permafrost. China contains the largest area with alpine permafrost, including the Qinghai-Tibet Plateau and portions of the Himalayas, Tien Shan, and Karakoram Mountains, at 2.1 million km^2, which constitutes 45 % of the total area with alpine permafrost (Table 10.1, Fig. 10.1). Russia contains the next largest area of mountain permafrost, including the Caucasus, Urals, Sayan–Stoblovoi Mountains and portions of the Altai Mountains at 581,000 km^2, followed by Mongolia, Canada, and the USA. However, as will be seen, only half of the alpine permafrost area contains cryosols, because the active layer is deeper than 1 or 2 m in many of these soils.

10.2 Soil-Forming Factors

10.2.1 Climate, Permafrost, and Active-Layer Depths

The elevation at which mountain permafrost is reported in the literature ranges from as low as 500 m (by definition; Gorbunov 1978) in high-latitude environments such as Iceland and the subpolar portions of the Caucasus Mountains to >5,000 m in mid-latitude regions such as the central Andes, Himalayas, and the Qinghai-Tibet Plateau (Table 10.2). The active-layer thickness ranges from >0.5 m in high-latitude mountain environments such as Iceland or Greenland to more than 8 m in the Andes, European Alps, and Altai Mountains.

Permafrost may exist in mountains where the mean annual air temperature (MAAT) is as warm as 1.4 °C, but a value of −3 °C or lower is more typical of areas containing mountain permafrost (Lewkowicz and Ednie 2004; Etzelmüller et al. 2007; Gruber 2012). A MAAT of −5 or −6 °C may be necessary for the active layer to be within

J.G. Bockheim, *Cryopedology*, Progress in Soil Science,
DOI 10.1007/978-3-319-08485-5_10

Table 10.1 World distribution of permafrost and cryosols

Country	Mountains	Low-elevation permafrost (10^3 km^2)	Alpine permafrost (10^3 km^2)[a]	Total permafrost (10^3 km^2)[b]	Low-elevation cryosols (10^3 km^2)	Alpine cryosols (10^3 km^2)	Total cryosols (10^3 km^2)	References
Russia	Caucasus, Urals, Altai, Sayan-Stoblovoi	**10,387**	**581**	10,968	7,270	**1,500**	9,720	Jones et al. (2010) and Marchenko (Personal communication)
Canada		**5,621**	**410**	6,031	2,330	170	2,500	Smith and Veldhuis (2004) and Tarnocai and Bockheim (2011)
China	Himalayas, Tien Shan, Karakoram, Qilian	0	2,056	2,056	0	**320**	**320**	This study
USA	Brooks, Wrangell, Rocky	**775**	**357**	1,132	603	190	793	Soil Survey Staff (1999)
Mongolia	Altai	0	530	530	0	**64**	**64**	Maximovich (2004)
Greenland		297	0	*297*	250	0	250	Jones et al. (2010)
Iceland		**2**	**8**	*10*	0	2	2	Jones et al., (2010)
Svalbard		**5**	**19**	*24*	6	18	24	Jones et al. (2010)
Norway		0	84	84	0	0	0	Jones et al. (2010)
Sweden		0	59	59	0	0	0	Jones et al. (2010)
Finland		0	34	34	0	0	0	Jones et al. (2010)
Kyrgystan	Tien Shan	0	72	72	0	**11**	**11**	
India	Himalayas, Karakoram	0	58	58	0	**6**	**6**	
Tajikistan	Pamirs	0	54	54	0	**8**	**8**	
Pakistan	Himalayas, Karakoram, Hindu Kush	0	40	40	0	**5**	**5**	

Kazakhstan	Altai	0	39	39	0	**5**	**5**	
Afghanistan	Himalayas, Pamirs, Hindu Kush	0	27	27	0	**3**	**3**	
Argentina	Andes	0	30	30	0	**2**	**2**	
Chile	Andes	0	14	14	0	**1**	**1**	
Nepal	Himalayas	0	15.7	15.7	0	**2**	**2**	
Bhutan	Himalayas	0	3.1	3.1	0	**0.4**	**0.4**	
Italy	European Alps	0	5.1	5.1	0	**0.5**	**0.5**	
Switzerland	European Alps	0	4.5	4.5	0	**0.5**	**0.5**	
Austria	European Alps	0	3.6	3.6	0	**0.4**	**0.4**	
Georgia	Caucasus	0	4.7	4.7	0	**0.6**	**0.6**	
North Korea	Taebaek	0	4.7	4.7	0	**0.5**	**0.5**	
Uzbekistan	Tien Shan, Pamir, Altai	0	2.8	2.8	0	**0.3**	**0.3**	
Antarctica	Sør Rondane, Transantarctic, Prince Charles, Pensacola, Ellsworth, Antarctic Peninsula	**3**	**43**	46	**3**	**43**	46	
	Total	17,090	4,559	21,649	10,462	2,353	13,765	

Bold-face values are estimates; values in italics were obtained from references listed
[a]Haeberli et al. (1993)
[b]Gruber (2012)

Fig. 10.1 Alpine soils of the world underlain by permafrost. Numbered locations: *1*=Coast Range; *2*=Rocky Mtns. (USA); *3*=Rocky Mtns. (Canada); *4*=Brooks Range; *5*=Cascade Range (Canada, USA); *6*=Appalachian Mtns.; *7*=Andes Mtns.; *8*=Fennoscandian mtns.; *9*=Icelandic mtns; *10*=Greenlandic mtns; *11*=Svalbard mtns.; *12*=European Alps; *13*=Pyrenees; *14*=Carpathian Mtns.; *15*=Urals; *16*=Caucasus; *17*=Qinghai-Tibet Plateau; *18*=Altai Mtns.; *19*=Pamir-Tien Shan Mtns.; *20*=Yablonai-Sayan-Stanovoi Mtns., *21*=Japanese Alps; *22*=Southern Alps, New Zealand

1–2 m of the ground surface in alpine regions (Bockheim and Munroe 2014). At the other extreme, MAAT values as low as −10 °C have been recorded in mountains of Alaska, Fennoscandia, and the Qinghai-Tibet Plateau (Table 10.2). The mean annual precipitation of areas with mountain permafrost ranges from 250 mm/year for the Yukon Territory of Canada and parts of the central Asian mountains to over 2,000 mm/year in the European Alps. Annual snowfall ranges from a meter to more than 20 m, although wind redistribution can produce significant local differences in snow depth.

10.2.2 *Biota*

Many alpine plants are circumpolar, meaning that they occur throughout the Arctic and mountains at lower latitudes, including Europe, Asia, and North America. There are a large number of species that are circumpolar, including *Pedicularis verticillata, Myosotis alpestris, Pyrola grandiflora, Carex bigelowii, Vaccinium vitis-idaea, V. ulignosum, Arctostaphylos alpina, and Dryas octopetala* (Körner 1999). According to Bliss (1979) there are fewer vascular plants in the Southern

Table 10.2 Alpine cryosols

Area	Latitude	Longitude	Elev. (m)	Active-layer depth (m)	MAAT (°C)	Conditions	References
Confirmed							
Iceland	65 30N	16W	800–900	0.4–0.6	–5		Etzelmüller et al. (2007) and Arnalds (2008)
QTP	29–33N	87–95	4,100–5,100	1.2–3.6	–5	Alpine meadow, steppe	Wang et al. (2002, 2008)
QTP	29–33N	87–95	4,100–5,100	<2	–5	Swamp	Wang et al. (2002, 2008)
QTP, Wudaoliang	35N	93E	4,800–5,200	<1.0	–5.7	Cryosols	Dörfer et al. (2013)
QTP, Golmud-Lhasa transect	30–36N	91–101E	4,900	1–2	–4.2	Cryosols	Baumann et al. (2009)
Swiss Alps	47N	6.5E	2,959	<1.0			Mutter and Phillips (2012)
Swiss Alps	47N	6.5E	2,390–2,959	1.2–1.5			Mutter and Phillips (2012)
Ruby Range, Yukon	61 12N	138 19W	2,200–2,300	>0.5 to <1.5	~–12		Bonnaventure and Lewkowicz (2013)
Urals, Yugyd Va	66N	61E	>850	0.3–0.5	–7	Peaty gleyzems	Dymov et al. (2013)
Urals, Rai-Iz massif	66.9N	66.5E	800–900	0.3	–8.2	Late June	Lesovaya et al. (2012)
Tokinskii-Stanovik Ridge, Russia	56N	131E	>1,750		~–5		Chevychelov and Volotovskii (2001)
Mongolia, Sayan Mtns., Hovsgol	50N	11E	1,659–1,806	1–2	–4.5	Wet valleys	Etzelmüller et al. (2006)
Kluane Lake, Yukon	61N	139W	823–1,310	0.4–0.9	–9		Harris (1987)
Beartooth Mtns. WY	45N	110W	>3,050	<1.0		Bogs, peat mounds	Johnson and Billings (1962)
Central Khangai, Mongolia	47N	100E	>2,500	1.4–1.5	–12	Mucky humus	Krasnoshchekov (2010)
Tien Shan Mtns.	40–44N	690–95E	>3,000	0.5–0.7	–8		Ping et al. (2002) and Marchenko et al. (2007)
Brooks Range, AK	69N	147W	900–1,200	0.5	–6.4	Mid-July	Ugolini and Tedrow (1963)

(continued)

Table 10.2 (continued)

Area	Latitude	Longitude	Elev. (m)	Active-layer depth (m)	MAAT		References
					(°C)	Conditions	
Brooks Range, AK	68N	159W	800	0.3–1.0	−5.9	Late July	Munroe and Bockheim (2001)
Ungave-Labrador Penin., Canada	59N	65W	590–1,005	<2.0	−5.7	Imperfect drainage	Hendershot (1985)
Sierras de Alvear, Tierra del Fuego	54 40S	68 02W	775–1,077	1.5–2.0	−2		Valcárcel-Díaz et al. (2009)
Possible							
Altai Mtns., Russia	48–52N	83–91E	2,300		−4.8	Maximum elev. 4,500 m	Fukui et al. (2007)
No cryosols							
Pyrenees, Europe	42N	25E	>3,050	1.0–4.0	−6.8		Serrano et al. (2001)
Finnish Lapland mtns.	69N	26E	>410, <640		−4		Hort and Luoto (2009)
Swedish Lapland, Tarfalaryggen	68N	18E	1,540	1.5	−6		Harris et al. (2009)
Swiss Alps, Schilthorn	47N	07E	2,909	5.0	−4.3		Isaksen et al. (2001) and Harris et al. (2009)
Swiss Alps, Murtel-Corvatsch	46N	10E	2,670	3.4	−3		Isaksen et al. (2001) and Harris et al. (2009)
Swiss Alps, Juvvasshoe	46N	08E	1,894	2.4	−4.5		Isaksen et al. (2001), Harris et al. (2009)
N Norway, Stockhorn	62N	08E	3,410	3.5	−5.5		Isaksen et al. (2001) and Harris et al. (2009)
Himalayas, Nepal	28N	86E	3,860–4,830			No cryosols	Bäumler and Zech (1994)
Himalayas, Lhasa	30N	91E	4,200–4,600	>2.0	+1.3	No cryosols	Smith et al. (1999)
Andes, Venezeula	09N	71W	5,000				Mahaney et al. (2009)
Andes, Argentina	32–33S		<3,300	>5.5			Trombotto and Borzotta (2009)

Mongolia, Hovsgol	49–52N	989–102E	1,547–1,864	1.0–3.0	−3	Fine texture soils	Sharkhuu (2003) and Sharkhuu et al. (2007)
Mongolia, Hovsgol	49–52N	989–102E	1,547–1,864	4–6	−3	Coarse-textured soils	Sharkhuu (2003) and Sharkhuu et al. (2007)
Old Man Range, NZ	45S	169E	1,670				Billings and Mark (1961)
Tatra Mtns., Poland	49N	20E	<2,600		−4		Skiba (2007)
Ben Ohau, Mt. Cook, NZ	44S	170E					Birkeland (1984)
Rocky Mtns., AB, Canada	50N	115W	2,300–2,700	≥1.5			Harris and Brown (1982)
Rocky Mtns., AB, Canada	51N	116W	2,240–2,969		<0	No cryosols	Knapik et al. (1973)
Uinta Mtns., UT, USA	41N	110W	3,528–3,735		−2.5	No cryosols	Bockheim and Koerner (1997) and Munroe (2007)
W, N Caucasus	43N	45E	<3,100		−2.9	No cryosols	Molchanov (2008, 2010)
Carpathian Mtns., Czech Rep.	47N	25E	<1,750		<0		Pelíŝek (1973)
Swedish Lapland	68N	18E	<1,585		−2	No cryosols	Darmody et al. (2000)
N. Norway	61N	8E	1,170–1,455		−2	No cryosols	Darmody et al. (2005)
QTP, Huashixia	31N	35E	–	4–7	−4.1		Dörfer et al. (2013)
QTP, Damxung	30N	91E	4,400–5,300		1.5		Ohtsuka et al. (2008)
QTP, Golmud-Lhasa transect	30–36N	91–101E		1.3–3.5			Pang et al. (2009)
QTP, Golmud-Lhasa transect	32–36N	91–94	4,400–5,100	1.0–3.5	−1.8 to −7.1		Li et al. (2012)
QTP, Chang Tang	33–35N	79–85E	4,500–5,300	>2.0	−0.2		Wu et al. (2012)

Circumpolar Region than in the Northern Circumpolar Region, and the physiognomy major vegetation types differ markedly from between the two poles.

10.2.3 Parent Materials and Time

The most common parent materials in alpine cryosols are colluvium, till, gelifluction deposits, and peat. Although these materials are commonly of Holocene age, some mountains escaped glaciation during the Last Glacial Maximum and feature very old soils. An example is the Uinta Mountains of Utah (Munroe 2007).

10.3 Soil Properties

The properties of alpine cryosols are highly variable, much like Arctic and Antarctic cryosols. From the 23 pedons in Table 10.3, silt concentrations range from 10–70 % (average = 35 %) and clay concentrations range from 4–31 % (average = 15 %); the pH ranges from 3.9 in organic horizons to 8.8 in semiarid cryosols of central Asia (average = 5.8); organic C concentrations exceed 20 % in organic horizons but are generally low in mineral horizons (overall average = 9.1 %); and soils derived from limestone and basalt materials have a high base saturation and those from acid igneous materials have a low base saturation. An isotic mineral class is common in mountain soils with permafrost in many areas, including the Rocky Mountains (Bockheim and Munroe 2014).

10.4 Soil-Forming Processes

In the scheme of Bockheim and Gennadiyev (2000), the dominant soil-forming process in high-mountain environments with permafrost is cryoturbation. Other processes of importance include andisolization, melanization (humification), cambisolization, podzolization, paludization, and gleization.

Although not requiring permafrost, cryoturbation (frost-stirring), is a common process in permafrost-affected soils and is manifested by patterned ground on the land surface and irregular and broken horizons, organic matter accumulation on the permafrost table, oriented stones, and silt caps within the soil. Patterned ground is a common feature in high-mountain environments (Johnson and Billings 1962). In addition to high-latitude mountain environments, such as Iceland (Arnalds 2008), Svalbard (Kabala and Zapart 2012), and the Scandinavian Mountains (Darmody et al. 2000), cryoturbation has been reported in the central Rocky Mountains of the U.S.A. (Bockheim and Koerner 1997; Munroe 2007), the Swiss Alps (Zollinger et al. 2013), the Ural Mountains (Dymov et al. 2013), and the Qinghai Plateau (Smith et al. 1999).

Table 10.3 Analytical properties of alpine cryosols

Location	Horizon	Depth (cm)	Silt (%)	Clay (%)	pH	SOC (%)	C:N	CEC (cmol+/kg)	BS (%)	Fe_d (%)
Polar Urals; Typic Aquiturbels (Lesovaya et al. 2012)										
	Cjjg1	0–10	35.2	11.1	8.8	0.7		17.1	4	**2.06**
	Cjjg2	10–20	47.3	17.2	7.8	0.5		28.1	5	**1.34**
Polar urals; Typic Aquiturbels (Lesovaya et al. 2012)										
	A	0–4	20.8	7.8	7.2	16.3		25.5	26	**2.68**
	Cg1	4–15	23.3	6.8	7	4.7		27.6	14	**4.05**
	Cg2	15–30	48.6	17.4	8	1.9		24.2	4	**1.53**
Subpolar Ural Mtns.,	Pedon: 25–09; Typic Aquiturbels (Dymov et al. 2013)									
Russia	Oi	0–5			3.9	41.4	56	66		
	Ag	5–14	36	21	5.2	0.8	12	8.8		
	BCg	14–37	37	22	5.1	0.7	12	9.6		
	Cg	37	30	18	5.2	0.1	2	8.1		
Northern Rocky Mtns., Sapristels (Johnson and Billings 1962)										
	Oa1	2–5	45.7	4.2	6.4	10.5				
	Oa2	15–20	41.2	12	5.2	8.8				
	Oa3	36–46	34.5	10.5	6.2	1				
Northern Rocky Mtns., Haploturbels (Johnson and Billings 1962)										
	A	2–16	30.9	4.9	5.3	9.8				
	C	15–30	28.4	4.7	5.5	5.9				
Northern Rocky Mtns., Haploturbels (Johnson and Billings 1962)										
	C1	2–5	66.9	7.3	5.8	0.29				
	C2	10–15	65.6	7.6	5.6	0.29				
	C3	36–38	69.5	8.5	6.6	0				
Central Khangai, Mongolia, Haplorthels (Krasnoshchekov 2010)										
	O	0–1	nd	nd	5.6	45.2		24.5	60	
	A	1–19	34	17	5.3	10.6		23.4	50	
	C	19–25	20	11	5.8	1.4		3.1	78	
Central Khangai, Mongolia, Umbrorthels (Krasnoshchekov 2010)										
	Oa	0–15	16	7	4.7	28.5	17	25.1	52	
	AC	15–25	10	6	4.2	15.7	34	23.2	70	
Central Khangai, Mongolia, Histoturbels (Krasnoshchekov 2010)										
	Oi	0–6			5.4	45.3		40.8	57	
	Oe	6–20			5.8	38.8		57.5	67	
	Crg	25–36	46	22	5.3	2.3		18.7	56	

(continued)

Table 10.3 (continued)

Location	Horizon	Depth (cm)	Silt (%)	Clay (%)	pH	SOC (%)	C:N	CEC (cmol+/kg)	BS (%)	Fe_d (%)
	Cg	36–55	39	23	5.2	1.5		13.1	72	
Tien Shan Mtns., China, Ruptic-Histic Aquiturbels (Ping et al. 2002)								XJ004		
	Oa	0–18			5.2					
	Bg	18–35			5.9					1.1
	Bg/Oajj	35–49			6.4					1.5
	Bw	55–82			6.7					1.4
	Oa/Bgjj	82–103			6.4					1.1
	C	103–140			6.8					1.2
	Cf	140–170			6.7					1.1
Tien Shan Mtns., China, Typic Hemistels (Ping et al. 2002)								XJ003		
	Oa	0–8			6.7					1.3
	Oe1	8–65			6					1.2
	Oe2	65–108			5.8					0.7
	Of	108–120			6.1					0.7
Ungava-Labrador Peninsula, Quebec, Canada; Typic Haploturbels (Hendershot 1985)										
	Cjj1	0–10			4.56	0.33				
	Cjj2	10–20			4.55	0.35				
	Cjj3	20–40			4.56	0.45				
	Cjj4	40–60			4.55	0.47				
Ungava-Labrador Peninsula, Quebec, Canada; Typic Haploturbels (Hendershot 1985)										
	A	0–2.5			3.9	4.14				
	Cjj1	2.5–12			4.87	0.74				
	Cjj2	12–33			5.55	0.3				
	Cjj3	33–55			5.25	0.71				
Ungava-Labrador Peninsula, Quebec, Canada; Typic Haploturbels (Hendershot 1985)										
	A	0–1.5			3.87	4.87				
	Cjj1	1.5–10			4.72	0.32				
	Cjj2	10–20			4.84	0.26				
	Cjj3	20–40			5.29	0.21				
Ungava-Labrador Peninsula, Quebec, Canada; Typic Umbriturbels (Hendershot 1985)										
	A1	0–4\			3.86	6.6				
	A2	4–10			4.41	1.52				
	Bwjj1	10–21			4.44	1.91				
	Bwjj2	21–35			4.54	2.15				
	Bwjj3	0–26			4.54	2.07				
	C	35–42			4.78	0.66				
Brooks Range, Alaska, USA; Lithic Haplorthels (Ugolini and Tedrow 1963)										
	Oa	0–10			7.5	23.9	18			

(continued)

Table 10.3 (continued)

Location	Horizon	Depth (cm)	Silt (%)	Clay (%)	pH	SOC (%)	C:N	CEC (cmol+/kg)	BS (%)	Fe_d (%)
	A	10–20			7.7	8.7	22			
	ACr	20–30			7.9	6.5	32			
Brooks Range, Alaska, USA; Lithic Sapristels (Ugolini and Tedrow 1963)										
	Oe1	0–10			6.4	31.3	15			
	Oe2	10–18			6.5	25.8	12			
	Oa1	18–28			7.6	24.5	12			
	Oa2	28–38			7.5	21.8	12			
Tibet Plateau; Typic Histoturbels (Baumann et al. 2009)					($CaCl_2$)					
	Oa	0–30			6.8		14			
	2BgCr	30–49			7.1		11			
	2C	49–80			7		8			
Tibet Plateau; Typic Umbrorthels (Baumann et al. 2009)										
	A	0–25			6.7		14			
	Bw	25–48			6.7		11			
	C	48–68			6.9		8			
	Cf	68–150			7		7			
Tokinskii Stanovik Ridge, Yakutia, Russia; Lithic Umbrorthels (Chevychelov and Volotovskii 2001)										
	Oi	0–2			5.3	54.3	80	40.1	19	
	A	2–10	10.5	5.2	5.2	9.6	21	9.4	30	**0.9**
	BhC	10–18	13	8.6	5	14.8	20	15.5	28	**0.71**
Galbraith-Toolik Lakes, Alaska; Typic Aquorthels (Munroe and Bockheim 2001)										
	Oe	0–2	37.5	27	5.05	2.8				
	Bw	2–15	43.2	26.6	5.1	2.7				
	Bg	15–38	49.5	17.6	5.66	5.6				
	BCg	38–60								
	Cgf	60								
Galbraith-Toolik Lakes, Alaska; Typic Mollorthels (Munroe and Bockheim 2001)										
	Oi	0–1								
	Oa	1–5								
	Bw1	5–10								
	Bw2	10–40	32.3	21.7	5.15	1.9				
	Bg	40–75	31	31	5.66	6.6				
	Oab	75–77								
	C	77–100	10.4	15.4	5.69	2.2				
Galbraith-Toolik Lakes, Alaska; Typic Histoturbels (Munroe and Bockheim 2001)										
	Oi	0–8								
	Oe	8–20								
	Oa	20–24								

(continued)

Table 10.3 (continued)

Location	Horizon	Depth (cm)	Silt (%)	Clay (%)	pH	SOC (%)	C:N	CEC (cmol+/kg)	BS (%)	Fe_d (%)
	Bwjj	24–36	39.1	24.9	6.13	4.6				
	Bgjj	36–50	35.4	24.5	6.2	2.6				
	Cg	50	38	26.5	6.86	2.2				

Andisolization refers to the formation of amorphous minerals (allophone) from weathering of volcanic ash and other silica-rich materials, a process that is especially prevalent in high-precipitation environments (Parfitt et al. 1983). High-mountain soils often have an isotic mineral class (Bockheim and Munroe 2014). These findings suggest that the formation of amorphous minerals common to high-mountain soils is favored by abundant soil moisture, a cold soil-temperature regime, and siliceous parent materials.

Melanization refers to the accumulation of well-humified organic matter within the upper mineral soil. This process is evidenced in high-mountains soils by the presence of A horizons (Table 10.3) and comparatively high SOC densities (Bockheim and Munroe 2014). Cambisolization leads to the formation of weakly developed Bw (cambic) horizons. This process is pervasive in mountain environments with permafrost (Table 10.3).

Podzolization is a complex collection of processes that includes eluviation of base cations, weathering transformation of Fe and Al compounds, mobilization of Fe and Al in surface horizons, and transport of these compounds to the spodic horizon as Fe and Al complexes with fulvic acids and other complex polyaromatic compounds. This process occurs in many high-mountain environments throughout the world, particularly in subalpine areas with a humid climate, ericaceous or coniferous vegetation, and siliceous parent materials (Burns 1990; Skiba 2007).

Gleization refers to redoximorphic features such as mottling and gleying that result from aquic conditions; this occurs in most mountain ranges, especially in bedrock depressions. Paludization refers to the accumulation of histic materials.

10.5 Soil Classification and Distribution

Fourteen (61 %) of the 23 pedons in Table 10.3 are Turbels, 7 (30 %) are Orthels, and 2 (10 %) are Histels. In the Arctic 64 % of the cryosols are Turbels (Chap. 8). Alpine cryosols occur primarily in the mountains of Arctic regions, such as in Iceland, the Brooks Range in northern Alaska, the cordillera of the northern Yukon Territory in Canada, and the subpolar Urals of Russia. Cryosols also occur in the mountains of central Asia, including above 5,000 m on the northern Qinghai-Tibet Plateau, above 3,000 m in the Tien Shan Mountains, above 2,500 m in the Khangai Mountains of Mongolia, and above 1,700 m in Sayan–Stolbovoi Mountains of Russia (Table 10.2).

Alpine cryosols comprise 2.3 million km^2, 17 % of world cryosol total (Table 10.1). These data suggest that only 52 % of the area mapped as containing alpine permafrost has permafrost within the upper 1–2 m. Although China contains 2.1 million km^2 of permafrost, only 320,000 km^2 feature cryosols.

Alpine soils described in the literature that have permafrost with active layers in excess of 1–2 m occur in seven other orders in *Soil Taxonomy*, including (from most to least prevalent) Inceptisols, Entisols, Spodosols, Histosols, Alfisols, Mollisols, and Andisols (Bockheim and Munroe 2014) (Fig. 10.2).

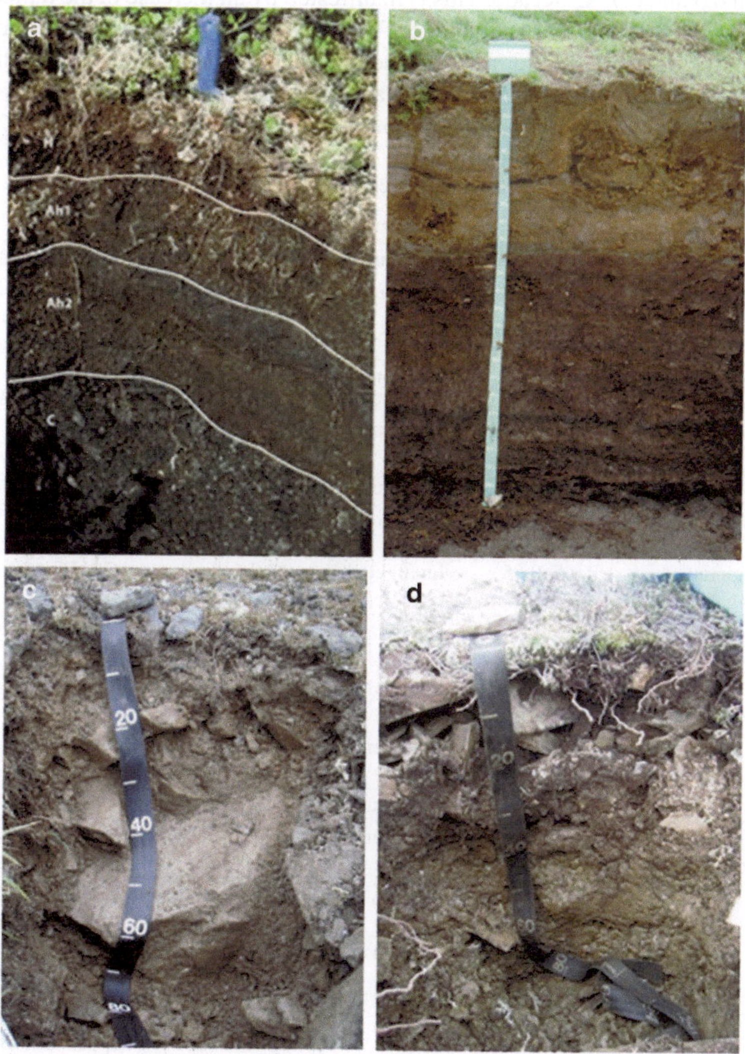

Fig. 10.2 Alpine soils with permafrost: (**a**) Richardson Mountains, Yukon Territory, Canada (P. Sanborn photo); (**b**) Cryoturbated Andisol, Iceland (http://www.rala.is/andisol); (**c**) Atigun Pass, Brooks Range, Alaska (NRCS photo); and (**d**) Eagle Summit, Alaska (NRCS photo)

10.6 Comparison of Alpine and Arctic Cryosols

Whereas alpine cryosols cover 2.3 million km^2, arctic cryosols have an area of 10.5 million km^2 (Table 10.4). From limited data (Bockheim and Munroe 2014), alpine cryosols contain an organic C density of 3.5–20.6 kg/m^2 to 1 m, where Arctic cryosols commonly have a range between 32 and 70 kg/m^2. The range in active-layer depths is similar for the two groups of cryosols. Both groups of cryosols have experienced increases in air temperature over the past several decades, commonly between 0.3 and 0.7 °C/decade. This is manifested by an increase in the temperature at the top of the permafrost table (TTOP) of between 0.1 and 0.7 °C/decade. Warming of the atmosphere has increased the active-layer depth on the Qinghai-Tibet Plateau by 1.3 cm/year, but has had no measureable impact on active-layer depth in the continuous permafrost region of the Arctic.

Table 10.4 A comparison of alpine and arctic cryosols

Parameter	Alpine	References	Polar	Reference
Area (10^6 km^2)	2.3	This study	10.5	Tarnocai et al. (2009)
SOC density (kg/m^2 to 1 m)	3.5–20.6	This study	32–70	Tarnocai et al. (2009) and Bockheim and Munroe (2014)
Patterned ground and/or cryoturbation (%)	5–19	Johnson and Billings (1962) and Feuillet (2011)	10–57	Bockheim et al. (1998) and Tarnocai et al. (1993)
Active-layer depth (m)	0.3–2.0	This study	0.3–2.0	Mazhitova et al. (2004) and Tarnocai et al. (2004)
Air temperature warming (°C/decade)	0.5–0.55 (Altai)	Fukui et al. (2007)	0.6 (arctic-wide)	CRUTEM 3v
	0.8 (QTP)	Li et al. (2012)	0.7 (W. Ant. Pen.)	Turner et al. (2009)
	1.0 (Front Range, CO)	Leopold et al. (2010)		
	0.06–0.3 (Tien Shan)	Marchenko et al. (2007)		
	0.6 (Altai)	Sharkhuu (2003)		
TTOP warming (°C/decade)	0.1–0.2 (Tien Shan)	Marchenko et al. (2007)	0.3–0.7 (N. Ak)	Osterkamp (2007)
	0.1–0.4 (QTP)	Li et al. (2012)	1.0 (NVL, Ant.)	Guglielmin and Cannone (2011)
	0.4–0.7 (Svalbard)	Isaksen et al. (2007)		
Active-layer depth (cm/year)	1.3 (QTP)	Li et al. (2012)	0 (NE Greenland)	Christiansen (2004)
			0 (Eur. Russia)	Mazhitova et al. (2004)
			0 (Arctic Canada)	Tarnocai et al. (2004)

10.7 Summary

The total area of alpine permafrost may be 3.6 million km^2 but there are only 2.3 million km^2 of alpine cryosols. A mean annual air temperature of −5 or −6 °C may be necessary for the active layer to be within 1–2 m of the ground surface in alpine regions. Many alpine plants are circumpolar, meaning that they occur throughout the Arctic and mountains at lower latitudes in the Northern Hemisphere. The properties of alpine cryosols are highly variable, much like Arctic and Antarctic cryosols. The dominant soil-forming process in alpine cryosols is cryoturbation; other processes of importance include andisolization, melanization (humification), cambisolization, podzolization, paludization, and gleization. Alpine cryosols occur primarily in the mountains of Arctic regions, but they also occur at high elevations in the mountains of central Asia. Alpine cryosols are comparable to Arctic cryosols, except that they contain considerably less organic C.

References

Arnalds O (2008) Soils of Iceland. Jökull 58:409–421

Baumann F, He J-S, Schmidt K, Kühn P, Scholten T (2009) Pedogenesis, permafrost, and soil moisture as controlling factors for soil nitrogen and carbon contents across the Tibetan Plateau. Glob Chang Biol 15:3001–3017

Bäumler R, Zech W (1994) Soils of the high mountain region of eastern Nepal: classification, distribution and soil forming processes. Catena 22:85–103

Billings WD, Mark AF (1961) Interactions between alpine tundra vegetation and patterned ground in the mountains of southern New Zealand. Ecology 42(1):18–31

Birkeland PW (1984) Holocene soil chronofunctions, Southern Alps, New Zealand. Geoderma 34:115–134

Bliss LC (1979) Vascular plant vegetation of the Southern Circumpolar Region in relation to Antarctic, alpine and arctic vegetation. Can J Bot 57:2167–2178

Bockheim JG, Gennadiyev AN (2000) The role of soil-forming processes in the definition of taxa in soil taxonomy and the world reference base. Geoderma 95:53–72

Bockheim JG, Koerner D (1997) Pedogenesis in alpine ecosystems of the eastern Uinta Mountains, Utah. Arctic Alp Res 29:164–172

Bockheim JG, Munroe JS (2014) Organic carbon pools and genesis of alpine soils with permafrost: a review. Arctic Antarct Alp Res 46 (in press)

Bockheim JG, Walker DA, Everett LR (1998) Soil carbon distribution in nonacidic and acidic tundra of arctic Alaska. In: Lal R, Kimble JM, Follett RF, Stewart BA (eds) Soil processes and the carbon cycle. CRC Press, Boca Raton, pp 143–155

Bonnaventure PP, Lewkowicz AG (2013) Mountain permafrost probability mapping using the BTS method in two climatically dissimilar locations, northwest Canada. Can J Earth Sci 45:433–455

Burns SF (1990) Alpine Spodosols: Cryaquods, Cryohumods, Cryorthods, and Placaquods above treeline. In: Kimble JM, Yeck RD (eds) Proceedings of the fifth International Correlation Meeting (ISCOM): characterization, classification, and utilization of Spodosols. USDA Soil Conservation Service, Lincoln, pp 46–62

Chevychelov AP, Volotovskii KA (2001) Soils of alpine and subalpine vertical zones of the Tokinskii Stanovik Ridge. Eurasian Soil Sci 34:704–709

Christiansen HH (2004) Meteorological control on interannual spatial and temporal variations in snow cover and ground thawing in two northeast Greenlandic Circumpolar-Active-layer-Monitoring (CALM) sites. Permafr Periglac Process 15:155–169
Darmody RG, Thorn CE, Dixon JC, Schlyter P (2000) Soils and landscapes of Kärkevagge, Swedish Lapland. Soil Sci Soc Am J 64:1455–1466
Darmody RG, Allen CE, Thorn CE (2005) Soil topochronosequences at Storbreen, Jotunheimen, Norway. Soil Sci Soc Am J 69:1275–1287
Dörfer C, Kühn P, Baumann F, He J-S, Scholten T (2013) Soil organic carbon pools and stocks in permafrost-affected soils on the Tibetan Plateau. PLoS One 8:e57024. doi:10.1371/journal.pone.0057024
Dymov AA, Zhangurov EV, Startsev VV (2013) Soils of the northern part of the subpolar Urals: morphology, physicochemical properties, and carbon and nitrogen pools. Eurasian Soil Sci 46:459–467
Etzelmüller B, Heggem ESF, Sharkhuu N, Frauenfelder R, Kääb A, Goulden C (2006) Mountain permafrost distribution modelling using a multi-criteria approach in the Hövsgöl area, northern Mongolia. Permafr Periglac Process 17:91–104
Etzelmüller B, Farbrot H, Gudmundsson A, Humlum O, Tveito OE, Björnsson H (2007) The regional distribution of mountain permafrost in Iceland. Permafr Periglac Process 18:185–199
Feuillet T (2011) Statistical analyses of active patterned ground occurrence in the Taillon Massif (Pyrénées, France/Spain). Permafr Periglac Process 22:228–238
Fukui K, Fujii Y, Mikhailov N, Ostanin O, Iwahana G (2007) The lower limit of mountain permafrost in the Russian Altai Mountains. Permafr Periglac Process 18:129–136
Gorbunov AP (1978) Permafrost investigations in high-mountain regions. Arctic Alp Res 10:283–294
Gruber S (2012) Derivation and analysis of a high-resolution estimate of global permafrost zonation. Cryosphere 6:221–233
Guglielmin M, Cannone N (2011) A permafrost warming in a cooling Antarctica? Climatic Change 111(2):177–195. doi:10.1007/s10584-011-0137-2
Haeberli W, Guodong C, Gorbunov AP, Harris SA (1993) Mountain permafrost and climate change. Permafr Periglac Process 4:165–174
Harris SA (1987) Altitude trends in permafrost active layer thickness, Kluane Lake, Yukon Territory. Arctic 40:179–183
Harris SA, Brown RJE (1982) Permafrost distribution along the Rocky Mountains in Alberta. In: Proceedings of the 4th Canadian permafrost conference, National Research Council of Canada, Calgary, 2–6 March, pp 59–67
Harris C, Arenson LU, Christiansen HH (19 co-authors) (2009) Permafrost and climate in Europe: monitoring and modeling thermal, geomorphological and geotechnical responses. Earth Sci Rev 92:117–171
Hendershot WH (1985) Comparison of Canadian and American classification systems for some Arctic soils of the Ungava-Labrador Peninsula. Can J Soil Sci 65:283–291
Hort J, Luoto M (2009) Interaction of geomorphic and ecologic features across altitudinal zones in a subarctic landscape. Geomorphology 112:324–333
Isaksen K, Holmlund P, Sollid JL, Harris C (2001) Three deep alpine-permafrost boreholes in Svalbard and Scandinavia. Permafr Periglac Process 12:13–25
Isaksen K, Sollid JL, Holmlund P, Harris C (2007) Recent warming of mountain permafrost in Svalbard and Scandinavia. J Geophys Res 112:F02S04. doi:10.1029/2006JF000522
Johnson PL, Billings WD (1962) The alpine vegetation of the Beartooth Plateau in relation to cryopedogenic processes and patterns. Ecol Monogr 32:105–135
Jones A, Montanaralla L, Stolbovoy V, Broll G, Tarnocai C, Spaargaren O, Ping C-L (2010) Soil atlas of the Northern circumpolar region. Joint Research Centre, European Commission, Rome
Kabala C, Zapart J (2012) Initial soil development and carbon accumulation on moraines of the rapidly retreating Werenskiold Glacier, SW Spitsbergen, Svalbard archipelago. Geoderma 175–176:9–20
Knapik LJ, Scotter GW, Pettapiece WW (1973) Alpine soil and plant community relationships of the Sunshine Area, Banff National Park. Arctic Alp Res 5:A161–A170

Körner C (1999) Alpine plant life: functional plant ecology of high mountain ecosystems. Springer, New York

Krasnoshchekov YN (2010) Soils and the soil cover of mountainous tundra and forest landscapes in the central Khangai of Mongolia. Eurasian Soil Sci 43:117–126

Leopold M, Voelkel J, Dethier D, Williams M, Caine N (2010) Mountain permafrost – a valid archive to study climate change? Examples from the Rocky Mountains Front Range of Colorado, USA. Nova Acta Leopoldina 112(384):281–289

Lesovaya SN, Goryachkin SV, Polekhovskii YS (2012) Soil formation and weathering on ultramafic rocks in the mountains tundra of the Rai-Iz massif, polar Urals. Eurasian Soil Sci 45:33–44

Lewkowicz AG, Ednie M (2004) Probability mapping of mountain permafrost using the BTS method, Wolf Creek, Yukon Territory, Canada. Permafr Periglac Process 15:67–80

Li R, Zhao L, Ding Y-J, Wu T-H, Xiao Y, Du E, Liu G-Y, Qiao Y-P (2012) Temporal and spatial variations of the active layer along the Qinghai-Tibet Highway in a permafrost region. Chin Sci Bull 57:4609–4616

Mahaney WC, Kalm V, Kapran B, Milner MW, Hancock RGV (2009) A soil chronosequence in Late Glacial and Neoglacial moraines, Humboldt Glacier, northwestern Venezuelen Andes. Geomorphology 109:236–245

Marchenko SS, Gorbunov AP, Romanovsky VE (2007) Permafrost warming in the Tien Shan Mountains, central Asia. Global Planet Change 56:311–327

Maximovich SV (2004) Geography and ecology of cryogenic soils of Mongolia. In: Kimble JM (ed) Cryosols; permafrost-affected soils. Springer, New York, pp 253–274

Mazhitova G, Malkova G, Chestnykh O, Zamolodchikov D (2004) Active-layer spatial and temporal variability at European Russian Circumpolar-Active-Layer-Monitoring (CALM) sites. Permafr Periglac Process 15:123–139

Molchanov EN (2008) Mountainous meadow Chernozem-like soils of high mountains in the North Caucasus region. Eurasian Soil Sci 41:1268–1281

Molchanov EN (2010) Mountain-meadow soils of the highlands in the western Caucasus. Eurasia Soil Sci 43:1330–1343

Munroe JS (2007) Properties of alpine soils associated with well-developed sorted polygons in the Uinta Mountains, Utah, U.S.A. Arctic Antarct Alp Res 39:578–591

Munroe JS, Bockheim JG (2001) Soil development in low-arctic tundra of the northern Brooks Range, Alaska, U.S.A. Arctic Antarct Alp Res 33:78–87

Mutter EZ, Phillips M (2012) Active layer characteristics at ten borehole sites in alpine permafrost terrain, Switzerland. Permafr Periglac Process 23:138–151

Ohtsuka T, Hirota M, Zhang X, Shimono A, Senga Y, Du M, Yonemura S, Kawashima S, Tang Y (2008) Soil organic carbon pools in alpine to nival zones along an altitudinal gradient (4400–5300 m) on the Tibetan Plateau. Polar Sci 2:277–285

Osterkamp TE (2007) Characteristics of the recent warming of permafrost in Alaska. J Geophys Res 112. doi:10.1029/2006JF000578

Pang Q, Cheng G, Li S, Zhang W (2009) Active layer thickness calculation over the Qinghai-Tibet Plateau. Cold Reg Sci Technol 57:23–28

Parfitt RL, Russell M, Orbell GE (1983) Weathering sequence of soils from volcanic ash involving allophone and halloysite, New Zealand. Geoderma 29:41–57

Pelíšek J (1973) Vertical soil zonality in the Carpathians of Czechoslovakia. Geoderma 9:193–211

Ping C-L, Zhao L, Wang S-L, Paetzold R, Kimble J, Bai-sheng YE (2002) Morphogenesis of Cryosols and associated soils in the alpine zone of Tienshan, West China. J Glaciol Geocryol 24:517–522

Serrano E, Agudo R, Delaloye R, González-Trueba JJ (2001) Permafrost distribution in the Posets massif, central Pyrenees. Nor J Geogr 55:245–252

Sharkhuu N (2003) Recent changes in the permafrost of Mongolia. In: Phillips M, Springman S, Arenson L (eds) Permafrost. Swets & Zeitlinger, Lisse, pp 1029–1034

Sharkhuu A, Sharkhuu N, Etzelmüller B, Heggem ESF, Nelson FE, Shiklomanov NI, Goulder CE, Brown J (2007) Permafrost monitoring in the Hovsgol mountain region, Mongolia. J Geophys Res 112. doi:10.1029/2006/F000543

Skiba M (2007) Clay mineral formation during podzolization in an alpine environment of the Tatra Mountains, Poland. Clays Clay Miner 55:618–634

Smith CAS, Veldhuis H (2004) Cryosols of the boreal, subarctic, and western cordillera regions of Canada. In: Kimble JM (ed) Cryosols; permafrost-affected soils. Springer, New York, pp 119–137

Smith CAS, Clark M, Broll G, Ping CL, Kimble JM, Luo G (1999) Characterization of selected soils from the Llasa region of Qinghai-Xizang Plateau, SW China. Permafr Periglac Process 10:211–222

Soil Survey Staff (1999) Soil taxonomy: a basic system of soil classification for making and interpreting soil surveys, 2nd edn, Agriculture handbook no. 436, USDA, Natural Resources Conservation Service, US Government of Printing Office, Washington, DC

Tarnocai C, Bockheim JG (2011) Cryosolic soils of Canada: genesis, distribution, and classification. Can J Soil Sci 91:749–762. doi:10.4141/CJSS10020

Tarnocai C, Nixon FM, Kutny L (2004) Circumpolar-Active-Layer-Monitoring (CALM) sites in the Mackenzie Valley, northwestern Canada. Permafr Periglac Process 15:141–153

Tarnocai C, Smith CAS, Fox CA (1993) International tour of permafrost affected soils: the Yukon and northwest territories of Canada. Centre for Land and Biological Resources Research, Research Branch, Agriculture Canada, Ottawa, 197 pp

Tarnocai C, Canadell JG, Schuur EAG, Kuhry P, Mazhitova G, Zimov S (2009) Soil organic carbon pools in the northern circumpolar permafrost region. Global Biogeochem Cycles 23. doi:10.1029/2008GB003327

Trombotto D, Borzotta E (2009) Indicators of present global warming through changes in active-layer thickness, estimation of thermal diffusivity and geomorphological observations in the Morenas Coloradas rockglacier, central Andes of Mendoza, Argentina. Cold Reg Sci Technol 55:321–330

Turner J, Bindschadler RA, Convey P, Di Prisco G, Fahrbach E, Gutt J, Hodgson DA, Mayewski PA, Summerhayes CP (2009) Antarctic climate change and the environment. Scientific Commissions on Antarctic Resolution, Cambridge

Ugolini FC, Tedrow JCF (1963) Soils of the Brooks Range: 3. Rendzina of the arctic. Soil Sci 96:121–127

Valcárcel-Díaz M, Carrera-Gómez P, Blanco-Chao R, Pérez-Alberti A (2009) Permafrost occurrence in southernmost South America (Sierra de Alvear, Tierra del Fuego, Argentina). In: Proceedings of the ninth international conference on permafrost, University of Alaska, Fairbanks, June 29–July 3, 2008, pp 1799–1802

Wang G, Cheng G, Shen Y (2002) Soil organic carbon pool of grassland soils on the Qinghai-Tibetan plateau and its global implication. Sci Total Environ 291(1–3):207–217

Wang G, Li Y, Wang Y, Wu Q (2008) Effects of permafrost thawing on vegetation and soil carbon pool losses on the Qinghai-Tibet Plateau, China. Geoderma 143:143–152

Wu X, Zhao L, Chen M, Fang H, Yue G, Chen J, Pang Q, Wang Z, Ding Y (2012) Soil organic carbon and its relationship to vegetation communities and soil properties across permafrost areas of the central western Qinghai-Tibet Plateau, China. Permafr Periglac Process 23:162–169

Zollinger B, Alewell C, Kneisel C, Meusburger K, Gärtner H, Brandová D, Ivy-Ochs S, Schmidt MWI, Egli M (2013) Effect of permafrost on the formation of soil organic carbon pools and their physical-chemical properties in the eastern Swiss Alps. Catena 110:70–85

Chapter 11
Cryosols and Earth-System Sciences

11.1 Introduction

Cryosols have been used extensively in glacial geomorphology, soil geomorphology, archaeology, and paleopedology. In this chapter we examine the use of cryosols in relative dating, correlation of glacial deposits, glacier dynamics, reconstruction of past environments, existence of former occupation sites, and other uses.

11.2 Cryosols and Relative Dating

A soil chronosequence is an array of related soils in a geographic area that differs primarily from the soil-forming factor, time; a chronofunction is the mathematical solution of the relationship (Jenny 1941):

$$S = f(t)cl, o, r, p$$

where the soil (S) and the properties that define it are functions of time (t), with the variables of climate (cl), organisms (o), relief (r), and parent material (p) remaining relatively constant. Using data from 32 chronosequences from 27 areas contained in the published literature, Bockheim (1980) showed that a single logarithmic model, $Y = a + b\ 10X$, yielded the highest correlation coefficients, when soil property, (Y), was correlated with time, (X), using linear regression techniques. He later sampled soils from 18 chronosequences in central and southern Victoria Land, 10 of which are in the MDVs (Fig. 11.1). For all of the chronosequences, there were highly significant correlations between time and soil properties, including depths of staining, maximum color development equivalence, visible salts, coherence, and ghosts (Fig. 11.2). Climate plays an interacting role in the slope of the regression lines relating soil property to time. For example, the profile accumulation of

J.G. Bockheim, *Cryopedology*, Progress in Soil Science,
DOI 10.1007/978-3-319-08485-5_11

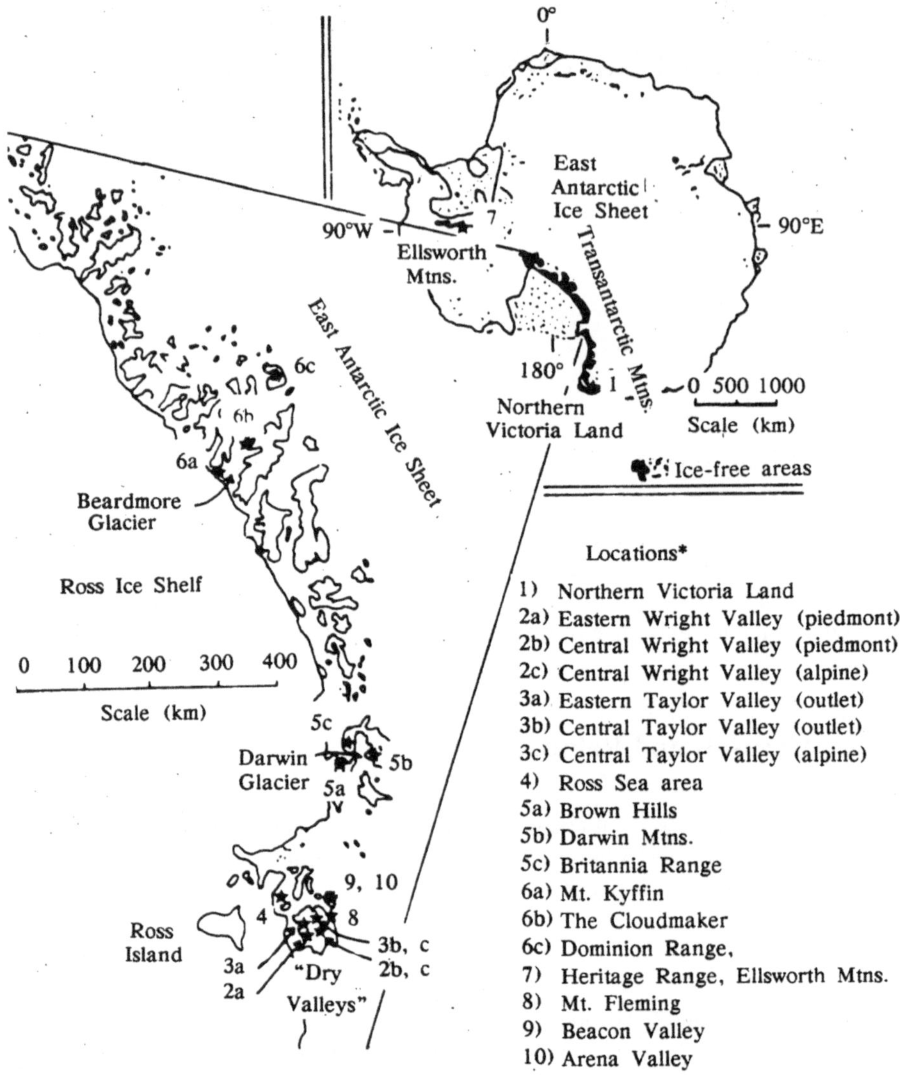

Fig. 11.1 Soil chronosequences of the Transantarctic Mountains (Bockheim and Wilson 1993)

soluble salts (to a depth of 70 cm) was greatest in xerous soils, followed by ultraxerous soils, with the least amounts in subxerous soils along the coast (Fig. 11.3).

Changes in morphological soil properties with time are readily visible in the chronosequence from lower Wright Valley (Fig. 11.4). The figure depicts soils on a Late Glacial Maximum surface (A), mid to late Quaternary (H1) aged hummocky drift (B), early Quaternary (Wright) drift (C), Pliocene-aged Valkyrie drift (D), Pliocene or older Alpine IV drift (E), and a strongly weathered soil below the white 3.9 Ma

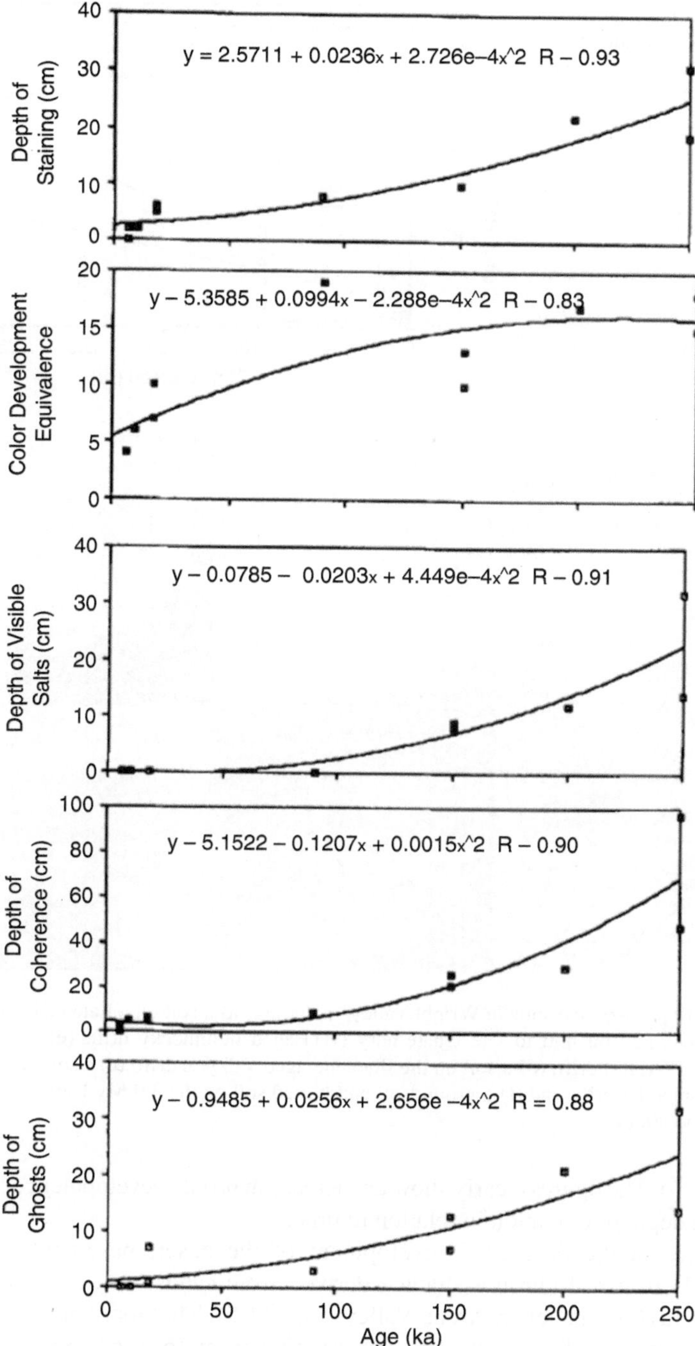

Fig. 11.2 Examples of soil chronofunctions from the central and southern Transantarctic Mountains (Bockheim 1990)

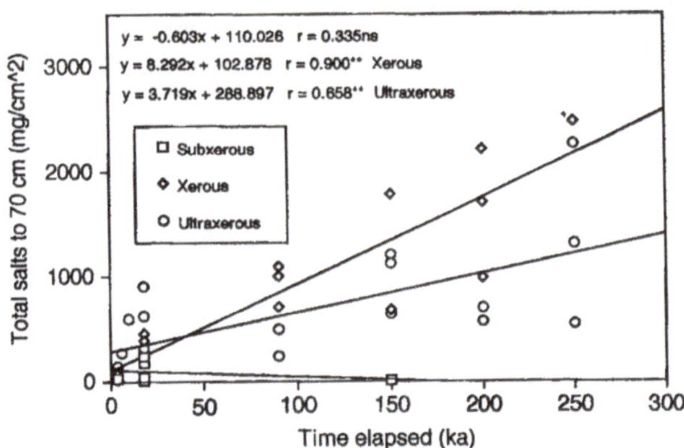

Fig. 11.3 Profile content of soluble salts in relation to time for three climatic zones in the central Transantarctic Mountains (Bockheim and Wilson 1993)

Fig. 11.4 Representative soils in Wright Valley, including (**a**) a soil on a Late Glacial Maximum surface, (**b**) a soil on mid to late Quaternary (H1) aged hummocky drift, (**c**) a soil on early Quaternary (Wright) drift, (**d**) a soil on the Pliocene-aged Valkyrie drift, (**e**) a soil on the Pliocene or older Alpine IV drift, and (**f**) a groundsoil and buried soil on the 3.9 My Hart Ash (Bockheim and McLeod 2006)

Hart Ash (F). The figures clearly show an increase in profile development of cohesion, salts, and depth of oxidation in relation to time.

Changes in the degree of development of the desert pavement are readily observable in a soil chronosequence derived from sandstone and dolerite drifts from the Taylor Glacier in Arena Valley (Fig. 11.5). The dominant size range of clasts decreases with time of exposure, ranging from 16 to 64 mm on Holocene and late Quaternary surfaces (A) to 8–16 mm on surfaces of middle Quaternary

Fig. 11.5 A chronosequence of desert pavements derived from sandstone and dolerite drifts from the Taylor Glacier in Arena Valley: (**a**) Taylor 2 drift (pedon 76–38); (**b**) Taylor 3 drift (pedon 86–23); (**c**) Taylor 4a drift (pedon 82–14); (**d**) Taylor 4b drift (pedon 76–29); (**e**) Altar drift (pedon 82–17); and (**f**) Arena drift (pedon 86–20) (Bockheim 2010a)

and older age (B, C, D, E, and F). The proportion of clasts with ventifaction increases progressively through time from 20 % on drifts of Holocene and late Quaternary age (A) to 35 % on Miocene-aged drifts (E, F). Desert varnish forms rapidly, especially on dolerite clasts, with nearly 100 % cover on surfaces of early Quaternary and older age. Macropitting occurs only on clasts that have been exposed since the Miocene (E, F).

The morphology of patterned ground changes through time as ice within polygon fissures sublimates (Fig. 11.6). In central Beacon Valley images obtained from a digital elevation model show the increasingly diffuse expression of high-center, sand-wedge polygons with time on three drift sheets on the valley floor, including regular pentagonal and hexagonal polygons on Taylor II drift (A), poorly expressed polygons on Taylor III drift (B), and diffuse polygons on Taylor IV drift (C). The apparent lineations on Taylor III and IV surfaces may reflect prevailing wind ablation from the southwest to northeast (Bockheim et al. 2009).

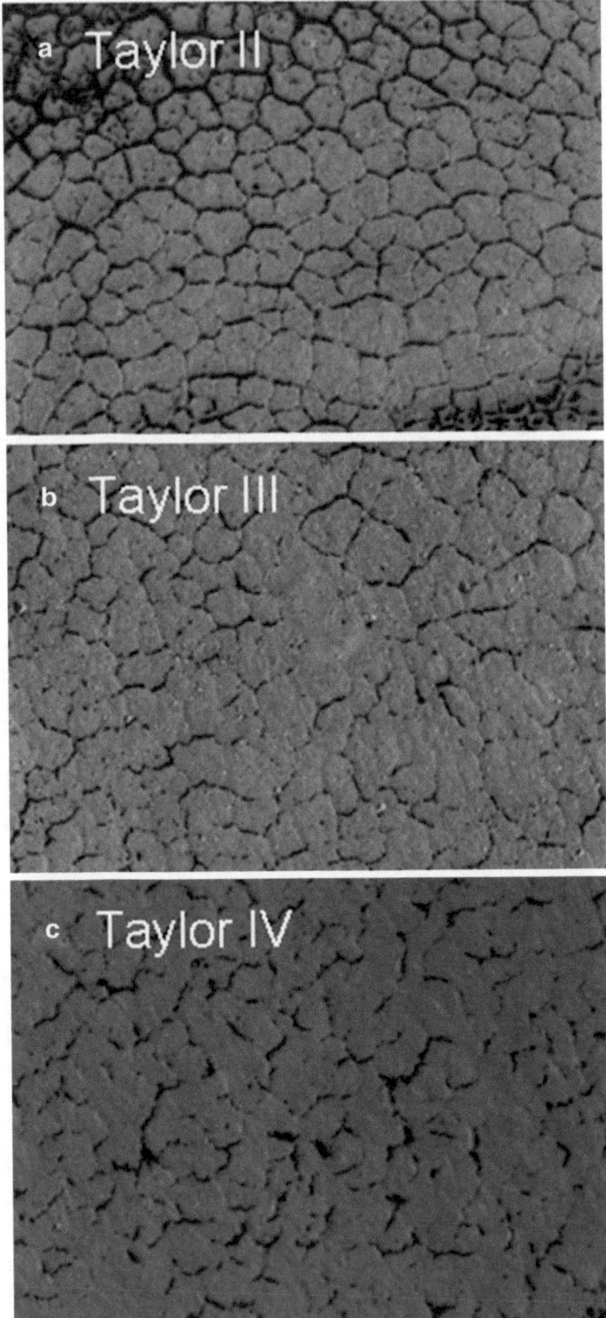

Fig. 11.6 Selected areas of patterned ground from the Beacon Valley digital elevation model for three drift sheets on the valley floor showing the increasingly diffuse expression of high-center, sand-wedge polygons with time in Beacon Valley: (**a**) Regular pentagonal and hexagonal polygons on Taylor II drift in lower Beacon Valley (from oblique aerial photo); (**b**) poorly expressed polygons on Taylor III drift in lower Beacon Valley; and (**c**) diffuse polygons on Taylor IV drift in central Beacon Valley. The apparent lineations on Taylor III and IV surfaces may reflect prevailing wind ablation from the southwest to northeast (Bockheim et al. 2009)

11.3 Cryosols and Correlation of Glacial Deposits

The fact that soils show a regular progression in development with time enables their use in correlating drifts between or among valleys affected by a similar glacial sequence. This has enabled us to develop a "master relative chronology" for the MDV (Table 11.1). The chronology is based on an examination of 431 sites on moraines with approximate ages that range from mid-Holocene to Miocene. The chronology enables investigators to estimate relative ages of landforms based on the properties listed, which include depths of staining, cohesion, visible salts, and ghosts (pseudomorphs), depth to ice-cemented permafrost, salt stage, weathering stage, thickness of salt pan, desert pavement development index, degree of patterned ground formation, and soil subgroup.

Table 11.2 shows a provisional correlation of drifts in the MDV based on data contained in Table 11.1. These data confirm that the outlet glaciers (Taylor, Wright Upper, Hatherton, Beardmore) and alpine glaciers (Wright Valley) have acted out-of-phase with grounded ice in the Ross Sea (Wilson Piedmont Glacier) and that outlet glaciers in the MDV behaved similarly in response to changes in climate that accompanied the glacial-interglacial cycles

11.4 Cryosols and Glacier Dynamics

Since the Pliocene most of the glaciers in the MDV have been cold-based (dry-based), meaning that they are frozen to their bed. These glaciers advance over frozen aprons at their termini so that they are able deposit drift with minimal impact to the underlying surface. Figure 11.7 provides evidence for overriding by cold-based glaciers in Arena Valley. The upper soil (above the diabase ventifact in the center of the image) is of Taylor IV age (>1.0 Ma, <7.4 Ma) and the buried soil is of Quartermain age (>11.3 Ma).

Bockheim (2010b) examined soil preservation and ventifact recycling from dry-based and wet-based glaciers at 609 sites in the central and southern TAM. Buried soils were most common from deposition by dry-based glaciers (44 of 51 pedons). Fifteen percent of the pedons contained recycled ventifacts in relict and buried soils that ranged from late Quaternary to Miocene in age, particularly in drift from dry-based glaciers (56 of 77 pedons). Overall 84 % of the buried soils and 78 % of the pedons with recycled ventifacts originated from dry-based glaciers. The proportion of soils with recycled clasts on a particular drift was greatest where the ratio of drift thickness to soil thickness ("recycling ratio") was the least.

Table 11.1 Master relative chronology for drifts in the McMurdo Dry Valleys

Drift unit	Approx. age	No. of sites	Staining	Max. CDE[a]	Coherence	Depth (cm) Vis. Salts	Ghosts	Ice cement	Salt stage[b]	Weathering stage[c]	Thickness salt pan (cm)	DPDI[d]	HCP index[e]	Soil sub-group[f]
					Taylor Valley (Bockheim et al. 2008)									
Alpine I	<3.7 Ky	3	0		11	0	0	16 (core)	0	1	0			GHt, THt
Ross Sea	12.4–23.8 Ky	7	0		28	5	0	34	1.1	1.6	0			THt
pre-Ross	>12.4–23.8 Ky	3	0		23	0	0	25	1	1.7	0			THt
Alpine II	113–120 Ky	32	10		21	7	8	>40	1.5	2.1	0			THt-TAo
Taylor II	113–120 Ky	14	3		18	3	6	>45	1.2	2.1	0	17		TAo-THt
Taylor III	208–375 Ky	27	18		>46	21	9	>85	2	2.7	0	17		TAo
Taylor IVa	1.6–2.1 My	28	38		>48	33	15	>44	3.4	4.2	4	22		TAo, SAo
Taylor IVb	2.7–3.5 My	18	33		>49	29	15	>46	2.8	4	4	21		TAo
Alpine III-IV	2.7–3.5 My	4	41		>47	41	14	>47	3.3	3.5	7			TAo
					Wright Valley (Bockheim and McLeod 2006)									
Alpine I	<3.7 Ky	5	0 (0)		5 (4)	0 (0)	0 (0)	12 (8)	0 (0)	1 (0)	0	16		GHt
Lacustrine	Holocene	1	0		5	1	0	50	0	1	0	nd		GHt

Brownworth	>49 Ka	5	0 (0)		23 (26)	3 (5)	0 (0)	48 (18)	1 (0)	1 (0)	0	nd		THt
Hummocky, H1	Late Quaternary	13	7 (8)		19 (22)	19 (21)	2 (5)	55 (18)	1 (1)	2 (1)	0	19		THt
Loke	Mid- to late Quaternary	2	0		31	17	0	33	1	2	0	23		THt
Hummocky, H2	Mid- to late Quaternary	7	33 (32)		>34	9 (19)	13 (7)	>97	2 (2)	3 (1)	0	20		TAo
Alpine II	<3.3 Ma	24	15 (7)		35 (23)	18 (18)	4 (5)	>65	2 (1)	3 (1)	0	18		TAo
Trilogy	Early-mid Quaternary	5	13 (22)		21 (29)	1 (1)	7 (12)	39 (7)	1 (1)	2 (2)		22		THt
Onyx	<3.3 Ma	10	29 (15)	10	45 (25)	32 (22)	13 (8)	>82	3 (2)	4 (1)	8	21		TAo, SAo
Wright	<3.4 Ma	11	>27	12	>30	22 (17)	16 (16)	>90	3 (2)	4 (1)	8	19		TAo, SAo
Valkyrie	Pliocene?	3	>44	12	>94	>40	39 (27)	>94	5 (1)	5 (0)	26	19		PsAo
Alpine III	<3.5 Ma	15	>43	12	>56	48 (15)	21 (19)	>103	4 (1)	6 (1)	16	23		PsAo
Alpine IV	>3.7 Ma	18	>55	12	>100	>63	19 (11)	>100	6 (1)	6 (0)	22	24		PsAo
Loop	Pliocene or Miocene	4	>60	13	>88	44 (9)	9 (4)	>88	5 (1)	6 (1)	12	22		PsAo
Peleus	>3.7 Ma, <5.4 Ma	6	24 (13)	6	>100	17 (4)	10 (5)	>100	4 (1)	5 (1)	8	20		TAo

(continued)

Table 11.1 (continued)

Drift unit	Approx. age	No. of sites	Staining	Max. CDE[a]	Coherence	Depth (cm) Vis. Salts	Ghosts	Ice cement	Salt stage[b]	Weathering stage[c]	Thickness salt pan (cm)	DPDI[d]	HCP index[e]	Soil sub-group[f]
					Arena & Beacon Valleys (Bockheim 2007)									
Taylor II	117 Ka	18	10		14	2	11	>28	1	2	0	19		TAo
post-Taylor III	>117 Ka, <200 Ka	1	17					>100	1	2	0	nd		TAo
Taylor III	200 Ka	15	25		52	19	16	>78	2.4	3	3	23		TAo
pre-Taylor III	>200 Ka	5	37		68	30	10	>60	4.2	4	15	nd		PnAo
Taylor IVa	>1.0 Ma, <2.2 Ma	13	42		38	28	29	>77	4.6	4.8	14	24		PnAo
Taylor IVb	>2.2 Ma, <7.4 Ma	13	42		39	33	27	>66	5	5.7	14	26		PnAo
Taylor IV, undiff.	>1.0 Ma, <7.4 Ma	10	28		47	16	12	>52	3.9	4.5	10	nd		TAo
Arena	>11.3 Ma	6	30		56	10	6	>75	3.2	4.2	6	26		TAo
Altar	>11.3 Ma	12	32		51	12	35	>47	3.9	4.6	10	26		TAo
Quartermain II	>11.3 Ma	2	35		>95	37	4	>90	4.5	5.5	14	26		TAo
		360												

					Victoria Valley System (Bockheim and McLeod 2013)									
Ice-cored	*[Alpine I]*	2	0	4	6	0	0	15	0.5	1.5	0		3.7	GHo-GHt
Packard	*[Taylor II]*	25	3	9	24	6	2	34	1.2	2.8	0		3.3	THo-THt
Vida	*[Taylor III]*	13	18	16	50	16	17	58	1.7	3.3	0		1.9	THo
Bull	*[Taylor IVb]*	14	32	15	64	28	21	>69	2.5	4.8	0		<0.5	TAo, SAo
Insel	*[Peleus, Arena, Altar]*	17	26	20	57	17	7	>57	2.7	4.7	0		<0.5	THo, TAo
		71												

[a]Color development index (Buntley and Westin, 1965)
[b]Bockheim (1997)
[c]Campbell and Claridge (1975)
[d]*DPDI* desert pavement development index (Bockheim 2010a)
[e]Maximum height of polygon, divided by mean width of contraction fissure
[f]*GHt* Glacic Haploturbels, *GHo* Glacic Haplorthels, *THt* Typic Haploturbels, *THo* Typic Haplorthels, *TAo* Typic Anhyorthels, *SAo* Salic Anhyorthels, *PsAo* Petrosalic Anhyorthels, *PnAo* Petronitric Anhyorthels

Table 11.2 Provisional correlation of glacial deposits[a] in the Transantarctic Mountains (Bockheim 2010a)

	Taylor V.	Wright V.	Wright V.	Arena V.	Beacon V.			
Geologic time scale	Taylor Gl.	Alpine	Wilson Pied. Gl.	Taylor Gl.	Taylor Gl.	Hatherton Glacier	Beardmore Glacier	Numerical Dating
Holocene		A1				Hatherton	Pl	3.7 Ky
Late quaternary			B			Br1, Br2	Be	10 Ky
	T2	A2a		T2	T2			117 Ky
			H1			D	M	
	T3	A2b		T3	T3			200 Ky
Middle quaternary			Loke, H2			I	pre-M	
Early quaternary	T4a			T4a	T4a			1.0–1.1 My
	T4b			T4b	T4b			1.1–2.2 My
Pliocene			O, W			pre-I	Do	<3.4 My
		A3						<3.5 My
			V					
		A4						>3.7 My
			P					
Miocene							S	7.7 My
				Al, Ar	Al			>11.3 My
References	Brook et al. (1993), Wilch et al. (1993), and Higgins et al. (2000)	Hall and Denton (2005)	Hall and Denton (2005)	Marchant et al. (1993)	Bockheim (2007)	Bockheim et al. (1989)	Denton et al. (1989) and Ackert and Kurz (2004)	

[a]Drift names: *A* Alpine, *Al* Altar, *Ar* Arena, *B* Brownworth, *Be* Beardmore, *Br* Britannia, *D* Danum, *Do* Dominion, *H* Hummocky, *Ha* Hatherton, *I* Isca, *L* Loke, *Lp* Loop, *M* Meyer, *P* Peleus, *Pl* Plunket, *O* Onyx, *Q* Quartermain, *S* Sirius, *T* Trilogy, *V* Valkyrie, *W* Wright

Fig. 11.7 Evidence for overriding by cold-based glaciers in Arena Valley. The upper soil (above the diabase ventifact in the center of the image) is of Taylor IV age (>1.0 Ma, <7.4 Ma) and the buried soil is of Quartermain age (>11.3 Ma)

These data illustrate the effectiveness of Antarctic dry-based glaciers in preserving underlying landforms and deposits, including soils. Moreover, the data imply that Antarctic glaciers have been recycling clasts for the past ca. 15 Ma. These findings have important implications in selecting surface boulders for cosmogenic dating.

11.5 Cryosols and Reconstruction of Past Environments

Cryosols have played an invaluable role in reconstructing past environments. A few examples will be given from the author's experience. Thaw lakes occur extensively in the coastal plain of the Eurasian and North American Arctic. About 20 % of the Arctic Coastal Plain of Alaska contains thaw lakes and another 50 % features drained thaw-lake basins (DTLBs) (Hinkel et al. 2003; Bockheim et al. 2004). The lakes and basins are elliptical with the long axis oriented a few degrees west of north and nearly perpendicular to the prevailing summer wind direction. It has long been known that these lakes undergo a natural cycle. They established a developmental sequence of DTLBs that included surface organic thickness, ice content of the uppermost permafrost, decomposition stage of organic matter, frost polygon form, great soil group, and vegetational parameters. The lowermost organic matter on the surface of the DTLBs was radiocarbon dated. Examples of the four age classes of DTLBs are shown in Fig. 11.8. The study validated that the thaw lake cycle occurs in response to climatic cooling and has been operating on the Barrow Peninsula for at least 5,500 year.

Many soils in the arctic contain high amounts of soil organic C in the transition layer, reflecting warmer periods when the thaw depth was deeper (Bockheim and Hinkel 2007). They showed that an average of 51.6 kg C m^{-3} was present in the

Fig. 11.8 Oblique photos of a chronosequence of drained thaw-lake basins near Barrow, Alaska: (**a**) <50 year, (**b**) 50–300 year, (**c**) 300–2,000 year, (**d**) 2,000–5,500 year (Hinkel et al. 2003)

Fig. 11.9 Organic carbon accumulated in the transition layer (from top of spade blade to halfway down where the gleyed material begins). The soil is a Ruptic-Histic Aquiturbel (J. Bockheim photo)

0–100 cm depth but that another 29 kg C m^{-3} occurred at the 100–200 cm depth, suggesting that at least 36 % of the soil organic C pool was below 100 cm (Fig. 11.9).

In Antarctica the evolution of soils reflects changes in climate and geologic conditions as the continent became separated and increasingly isolated from Gondwana. A greenhouse climate existed during the middle Paleozoic; and an ice-house climate began in the early Oligocene. In the Oligocene, Nothofagus–Podocarpaceae forests contained gelisols (Retallack et al. 2002). The climate of the Transantarctic Mountains has become increasingly hyperarid since the middle Miocene as reflected by the presence of soils derived from silt-rich till from a warm-based glacier (Bockheim 2013).

11.6 Cryosols in Archaeology

Humans have occupied the Arctic for 13,000 years (Buckland et al. 2011) and high mountains such as the Altai for at least 2,500 years (Epov et al. 2012). Although Antarctica has never had a native, permanent human population, early sealers and whales from the past 200 years have left an imprint (Villagran et al. 2013).

Cryosols have received less attention in archaeological studies than other soil orders. There are several reasons for this: the total number of people living above the Arctic Circle (66.56°N) is around four million in an area of over 12 million km^2, there are no permanent settlements above 78°N or in Antarctica, and the high alpine environments have had only sparse settlement, primarily in the Andes, Himalayas, and Altai.

Several examples will be given to illustrate the importance of permafrost in preserving artefacts and human remains and three case studies will be presented describing the use of soils in validating human occupation of a site in the Altai, the Canadian Arctic, and Antarctica. There is widespread concern that warming in the Arctic will destroy human occupation sites because of thawing of permafrost which preserves artefacts and human remains and rising sea levels which will flood coastal occupation sites (Blankholm 2009).

Derry et al. (1999) reported elevated levels of total N and P and lower pH values on Dorset/Thule occupation sites on Igloolik Island, Nunavut (69°22′N, 81°47′W) that were occupied for a 1,000 years prior to 1823. There are more than 1,000 frozen burial mounds on the Gorny Altai plateau at 2,500–2,600 m a.s.l. that date back to the Early Iron Age (2.5 kyr) (Epov et al. 2012). These mounds contain artefacts, mummies, and items made of textiles, wood, and leather that are preserved by permafrost. In the last example, Villagran et al. (2013) reported higher levels of P_2O_5, CaO, and total C on former sealing sites from the nineteenth century on Byers Peninsula (Livingston Island, Antarctica).

11.7 Cryosols as Extraterrestrial Analogs

Because of the hyperarid, hypothermal environments on Mars and Antarctica, cryosols and their accompanying landforms in the McMurdo Dry Valleys have been used as a Martian analog for more than 40 years (Morris et al. 1972; Anderson et al. 1972; Berkley and Drake 1981) (Fig. 11.10). Comparisons between Mars and Antarctica have included pitting of surface boulders (Head et al. 2011), the composition of oxidative weathering rinds on rocks (Salvatore et al. 2013), weathering products in soils such as zeolites (Berkley and Drake 1981), the morphology and chemistry of Miocene-aged paleosols in Antarctica (Mahaney et al. 2011), the composition of water and diagenetic minerals in permafrost (Dickinson and Rosen 2003; Heldmann et al. 2013), and patterned ground features (Mellon et al. 2009).

11.8 Cryosol Permafrost and Ancient Microorganisms

One of the more exciting discoveries in the cryosol regions has been the identification of microbial life in ancient permafrost. This work was championed by David Gilichinsky and his colleagues (1993) who isolated viable microbial cells from permafrost samples taken throughout the cold regions of the earth, including fungi, yeasts, and actinomycetes. The microbes were collected from permafrost ranging

Rock Weathering In Antarctica and Mars

Antarctica

Mars

Antarctica

Mars

Fig. 11.10 A comparison of landform surfaces in Antarctica and Mars (Source: NASA)

from 2.5 to 5 Ma in age from the Arctic, Antarctica, and the high mountains of Siberia. Samples collected from the Antarctic dry valleys were analyzed for biodiversity, state, and age (Gilichinsky et al. 2007).

11.9 Cryosols and High-Level Lakes in the McMurdo Dry Valleys

According to Hall et al. (2000), a high-water-level (336 m) lake, Glacial Lake Washburn, existed throughout Taylor Valley during the Last Glacial Maximum (LGM) and early Holocene, ca. 18.6–6.0 ka. They projected that this lake was 38 km^2 in area and had a maximum depth of 300 m. Hall et al. (2001) and Hall and Denton (2005) proposed the existence of Glacial Lake Wright, a high-water-level (550 m) lake, during the LGM and early Holocene, ca. 2.7–25.7 ka in Wright Valley. They proposed that this lake was 212 km^2 in area and had a maximum depth of 470 m. Hall et al. (2002) suggested that a 185-m deep lake may have engulfed most of the Victoria Valley system between 20 and 8.6 ka. The primary evidence for high-level lakes is the presence of deltas containing cyanobacterial mats that have been radiocarbon dated.

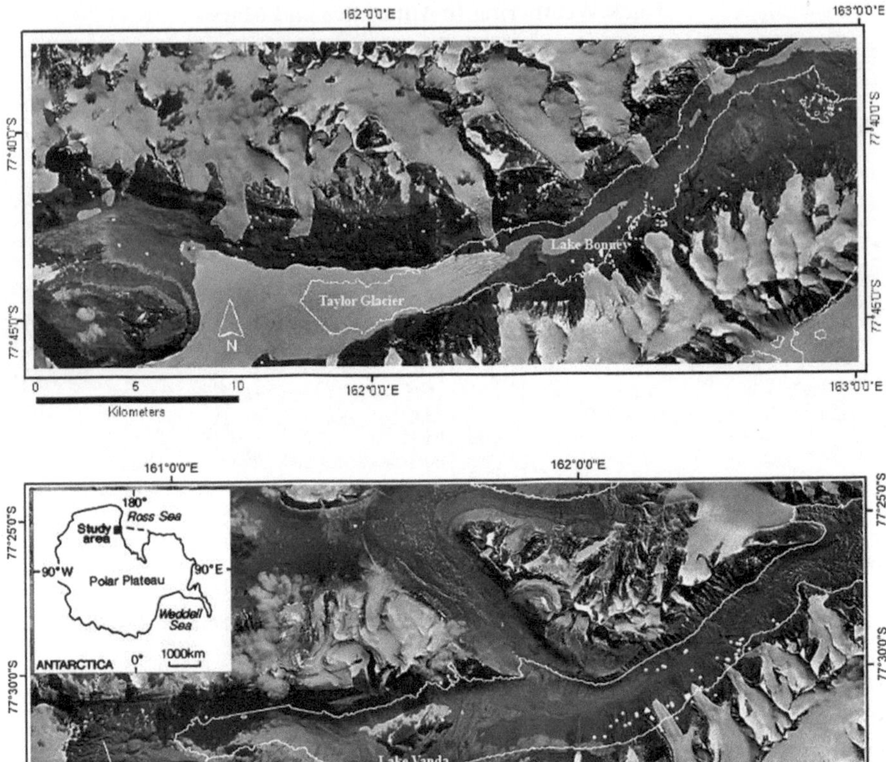

Fig. 11.11 Location of soil pits above and within the proposed upper elevations of Glacial Lakes Taylor and Wright by Hall et al. (2000, 2001). Maps from Bockheim et al. (2008)

Bockheim et al. (2008) and Bockheim and McLeod (2013) hypothesized that soils above the uppermost paleolake levels should be more strongly developed and contain more salts than soils below. In central Taylor and Wright Valleys, soils on equivalent-aged drifts above and below the conjectured upper limits of Glacial Lakes Washburn (336 m) and Wright (550 m), respectively, are all well developed with no appreciable differences in their properties (Fig. 11.11). Moreover, there were no significant differences in the slopes of regression equations relating soil property to age of the parent materials above and below the high-water lake levels (Fig. 11.12). Other than small alluvial fans with algae at all elevations, they found no evidence of former lake sediments nor did they find high-level strandlines except for strandlines on the north valley wall ca. 50 m above Lake Vanda, ice-shove features, or paleo-shore features. In Victoria Valley, a regression analysis of depth of

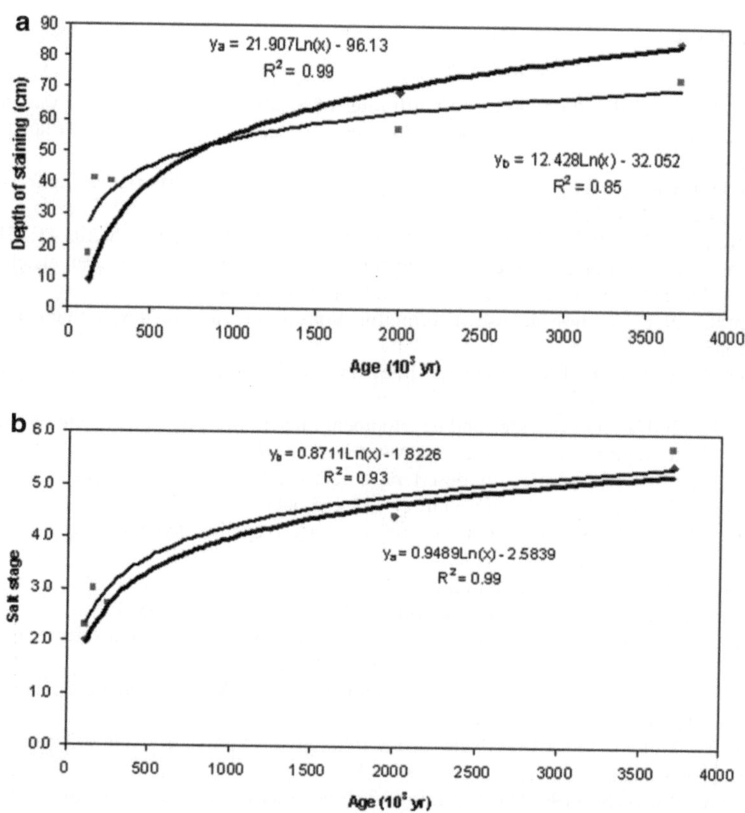

Fig. 11.12 Depth of staining (*top*) and salt stage for soils (**a**) (Ya) and (**b**) (Yb) the 550 m contour of Wright Valley (Bockheim et al. 2008)

visible salts against elevation yielded a very poor adjusted R^2 of 0.27 (Bockheim and McLeod 2013). They argued that lakes of the magnitude and duration proposed by Hall et al. (2002) would have dissolved and redistributed salts in the soils.

11.10 Summary

Cryosols have played an important role in understanding Earth's systems, including relative dating of soil parent materials, correlating geologic deposits, understanding glacier dynamics, reconstructing past environments, preservation of artefacts and microorganisms, detecting paleo-human occupation sites, predicting soils and geomorphic surfaces on extraterrestrial planets such as Mars, and determining whether or not high level lakes existed during the early Holocene in the McMurdo Dry Valleys.

References

Ackert RP Jr, Kurz MD (2004) Age and uplift rates of Sirius Group sediments in the Dominion Range, Antarctica, from surface exposure dating and geomorphology. Global Planet Change 42:207–225

Anderson DM, Gatto LW, Ugolini FC (1972) An Antarctic analog of Martian permafrost terrain. Antarct J US 7:114–115

Berkley JL, Drake MJ (1981) Weathering of Mars: Antarctic analog studies. Icarus 45:231–249

Blankholm HP (2009) Long-term research and cultural resource management strategies in light of climate change and human impact. Arctic Anthropol 46:17–24

Bockheim JG (1980) Solution and use of chronofunctions in studying soil development. Geoderma 23:71–85

Bockheim JG (1990) Soil development rates in the Transantarctic Mountains. Geoderma 47:59–77

Bockheim JG (2007) Soil processes and development rates in the Quartermain Mountains, upper Taylor Glacier region, Antarctica. Geogr Ann 89A:153–165

Bockheim JG (2010a) Evolution of desert pavements and the vesicular layer in soils of the Transantarctic Mountains. Geomorphology 118:433–443

Bockheim JG (2010b) Soil preservation and ventifact recycling from dry-based glaciers in Antarctica. Antarct Sci 22:409–417

Bockheim JG (2013) Soil formation in the Transantarctic Mountains from the middle Paleozoic to the Anthropocene. Palaeogeogr Palaeoclimatol Palaeoecol 381–382:98–109

Bockheim JG, Hinkel KM (2007) The importance of "deep" organic carbon in permafrost-affected soils of arctic Alaska. Soil Sci Soc Am J 71:1889–1892

Bockheim JG, McLeod M (2006) Soil formation in Wright Valley, Antarctica since the late Neogene. Geoderma 137:109–116

Bockheim JG, McLeod M (2013) Glacial geomorphology of the Victoria Valley system, Ross Sea Region, Antarctica. Geomorphology 193:14–24

Bockheim JG, Wilson SC (1993) Soil-forming rates and processes in cold desert soils of Antarctica. In: Gilichinsky DG (ed) Cryosols: the effects of cryogenesis on the processes and pecularities of soil formation. Proceedings of the 1st international conference on cryopedology. Russian Academy of Sciences, Pushchino, Moscow region, pp 42–55

Bockheim JG, Hinkel KM, Eisner WR, Dai XY (2004) Carbon pools and accumulation rates in an age-series of soils in drained thaw-lake basins, arctic Alaska. Soil Sci Soc Am J 68:697–704

Bockheim JG, Campbell IB, McLeod M (2008) Use of soil chronosequences for testing the existence of high-water-level lakes in the McMurdo Dry Valleys, Antarctica. Catena 74:144–152

Bockheim JG, Kurz MD, Soule SA, Burke A (2009) Genesis of active sand-filled polygons in lower and central Beacon Valley, Antarctica. Permafr Periglac Process 20:295–308

Brook EJ, Kurz MD, Ackert R Jr, Denton GH, Brown ET, Raisbeck GM, Yiou E (1993) Chronology of Taylor Glacier advances in Arena Valley, Antarctica, using in-situ cosmogenic 3He and 10Be. Quatern Res 39:11–23

Buckland PI, Eriksson E, Linderholm J, Viklund K, Engelmark R, Palm F, Svensson P, Buckland P, Panagiotakopulu E, Olofsson J (2011) Integrating human dimensions of Arctic palaeoenvironmental science: SEAD – the strategic environmental archaeology database. J Archaeol Sci 38:345–351

Denton GH, Bockheim JG, Wilson SC, Stuiver M (1989) Late Wisconsin and early Holocene glacial history, inner Ross Embayment, Antarctica. Quatern Res 31:151–182

Derry AM, Kevan PG, Rowley SDM (1999) Soil nutrients and vegetation characteristics of a Dorset/Thule site in the Canadian Arctic. Arctic 52:204–213

Dickinson WW, Rosen MR (2003) Antarctic permafrost: an analogue for water and diagenetic minerals on Mars. Geology 31:199–202

Epov MI, Balkov EV, Chemyakina MA, Manshtein AK, Manshtein YU, Napreev DV, Kovbasov KV (2012) Frozen mounds in Gorny Altai: geophysical and geochemical studies. Russ Geol Geophys 53:583–593

Gilichinsky DA, Wilson GS, Friedmann EI, 20 co-authors (2007) Microbial populations in Antarctic permafrost: biodiversity, state, age, and implication for astrobiology. Astrobiology 7:275–311

Hall BL, Denton GH (2005) Surficial geology and geomorphology of eastern and central Wright Valley, Antarctica. Geomorphology 64:25–65

Hall BL, Denton GH, Hendy CH (2000) Evidence from Taylor Valley for a grounded ice sheet in the Ross Sea, Antarctica. Geograf Annaler 82A:275–303

Hall BL, Denton GH, Overturf B (2001) Glacial Lake Wright, a high-level Antarctic lake during the LGM and early Holocene. Antarct Sci 13:53–60

Hall BL, Denton GH, Overturf B, Hendy CH (2002) Glacial Lake Victoria, a high-level Antarctic lake inferred from lacustrine deposits in Victoria Valley. J Quat Sci 17:697–706

Head JW, Kreslavsky MA, Marchant DR (2011) Pitted rock surfaces on Mars: a mechanism of formation by transient melting of snow and ice. J Geophys Res 116:E09007. doi:10.1029/2011JE003826

Heldmann JL, Pollard W, McKay CP, Marinova MM, Davila A, Williams KE, Lacelle D, Andersen DT (2013) the high elevation dry valleys in Antarctica as analog sites for subsurface ice on Mars. Planet Space Sci 85:53–58

Higgins SM, Hendy CH, Denton GH (2000) Geochronology of Bonney drift, Taylor Valley, Antarctica: evidence for interglacial expansions of Taylor Glacier. Geogr Ann 82A:391–409

Hinkel KM, Eisner WR, Bockheim JG, Nelson FE, Peterson KM, Dai X (2003) Spatial extent, age and carbon stocks in drained thaw lake basins on the Barrow Peninsula, Alaska. Arctic Antarct Alp Res 35:291–300

Jenny H (1941) Factors of soil formation. McGraw-Hill, NY

Mahaney WC, Hart KM, Dohm JM, Hancock RGV, Costa P, O'Reilly SS, Kelleher BP, Schwartz S, Lanson B (2011) Aluminum extracts in Antarctic paleosols: proxy data for organic compounds and bacteria and implications for Martian paleosols. Sedim Geol 237:84–94

Marchant DR, Denton GH, Swisher CC III (1993) Miocene-Pliocene Pleistocene glacial history of Arena Valley, Quartermain Mountains, Antarctica. Geogr Ann 75A:269–302

Mellon MT, Malin MC, Arvidson RE, Searls ML, Sizemore HG, Heet TL, Lemmon MT, Keller HU, Marshall J (2009) The periglacial landscape at the Phoenix land sites. J Geophys Res 114. doi:10.1029/2009JE003418

Morris EC, Holt HE, Mutch TA, Lindsay JG (1972) Mars analog studies in Wright and Victoria Valleys, Antarctica. Antarct J US 7:113–114

Retallack GJ, Krull ES, Bockheim JG (2002) New grounds for reassessing paleoclimate of the Sirius Group, Antarctica. J Geol Soc (Lond) 158:925–933

Salvatore MR, Mustard JF, Head JW, Cooper RF, Marchant DR, Wyatt MB (2013) Development of alteration rinds of oxidative weathering processes in Beacon Valley, Antarctica, and implications for Mars. Geochim Cosmochim Acta 115:137–161

Villagran XS, Schaefer CEGR, Ligouis B (2013) Living in the cold: geoarchaeology of sealing sites from Byers Peninsula (Livingston Island, Antarctica). Quat Int 315:184–199

Wilch TI, Denton GH, Lux DR, McIntosh WC (1993) Limited Pliocene glacier extent and surface uplift in middle Taylor Valley, Antarctica. Geogr Ann 75A:331–351

Chapter 12
Cryosols in a Changing Climate

12.1 Introduction

The high-latitude and high-elevation regions are undergoing more rapid and greater changes in climate than anywhere else on Earth. Annual air temperatures in alpine areas around the world have increased from 0.3 to 1.0 °C per decade for the past several decades (Table 12.1). According to CRUTEM, the arctic-wide increase in air temperature is 0.6, which is comparable to the 0.7 °C per decade value reported by Turner et al. (2009) for the western Antarctic Peninsula over the past several decades. To put these numbers in perspective, the world annual air temperature has increased 0.06 °C per decade since 1880 (IPCC 2013).

The reasons for the rapid increase in air temperature at the high latitudes and elevations pertain to the effect of atmospheric warming on snow cover and albedo. Let us now look at the impacts of this warming on the properties and distribution of cryosols.

12.2 Active-Layer Depths

Since the active-layer depth (ALD) is dependent on such factors as snow cover and air temperature, warming should result in an increase in ALD, especially in areas of sporadic or isolated permafrost but also along the southern boundary of discontinuous permafrost. There should be no change in ALD in the area of continuous permafrost because of the universally cold air and permafrost temperatures. This indeed appears to be the case. Based on data from CALM sites (Chap. 2), the active-layer has decreased 1.3 cm/year on the Qinghai-Tibet Plateau (QTP) and 1.4 cm/year in northern Sweden, areas of sporadic and discontinuous permafrost (Table 12.1). In contrast, there has been no statistically significant change in ALD in areas of continuous permafrost, such as northeast Greenland, European Russia, Arctic Canada, and northern Alaska.

J.G. Bockheim, *Cryopedology*, Progress in Soil Science,
DOI 10.1007/978-3-319-08485-5_12

Table 12.1 A comparison of geomorphology and soils of high-elevation and high-latitude environments

Parameter	Alpine	References	Polar	References
Area (10^6 km^2)	3.3	This study	22.8	Tarnocai et al. (2009)
SOC density (kg/m^2 to 1 m)	6.7–45	This study	32–70	Tarnocai et al. (2009)
SOC storage (Pg to 1 m)	55	This study	496	Tarnocai et al. 2009
Patterned ground and/or cryoturbation (%)	5–19	Johnson and Billings (1962) and Feuillet (2011)	10–57	Bockheim et al. (1998) and Tarnocai et al. (2009)
Active-layer depth (m)	2.0–8.0	This study	0.3–2.0	Mazhitova et al. (2004) and Tarnocai et al. (2004)
Air temperature warming (°C/decade)	0.5–0.55 (Altai)	Fukui et al. (2007)	0.6 (arctic-wide)	CRUTEM 3v
	0.8 (QTP)	Li et al. (2012)	0.7 (W. Ant. Pen.)	Turner et al. (2009)
	1.0 (Front Range, CO)	Leopold et al. (2010)		
	0.06–0.3 (Tien Shan)	Marchenko et al. (2007)		
	0.6 (Altai)	Sharkhuu (2003)		
TTOP warming (°C/decade)	0.1–0.2 (Tien Shan)	Marchenko et al. (2007)	0.3–0.7 (N. Ak)	Osterkamp (2007)
	0.1–0.4 (QTP)	Li et al. (2012)	1.0 (NVL, Ant.)	Guglielmin and Cannone (2011)
	0.4–0.7 (Svalbard)	Isaksen et al. (2007)		
Active-layer depth (cm/year)	1.3 (QTP)	Li et al. (2012)	0 (NE Greenland)	Christiansen (2004)
			0 (Eur. Russia)	Mazhitova et al. (2004)
			0 (Arctic Canada)	Tarnocai et al. (2004)
			1.4 (Sweden)	Christiansen et al. (2010)

12.3 Permafrost Temperatures

A critical measurement for showing any change in permafrost due to climate warming is the Mean Annual Ground Temperature (MAGT), usually taken at 10 m, or the Temperature at the Top of Permafrost (TTOP). From boreholes that are part of the

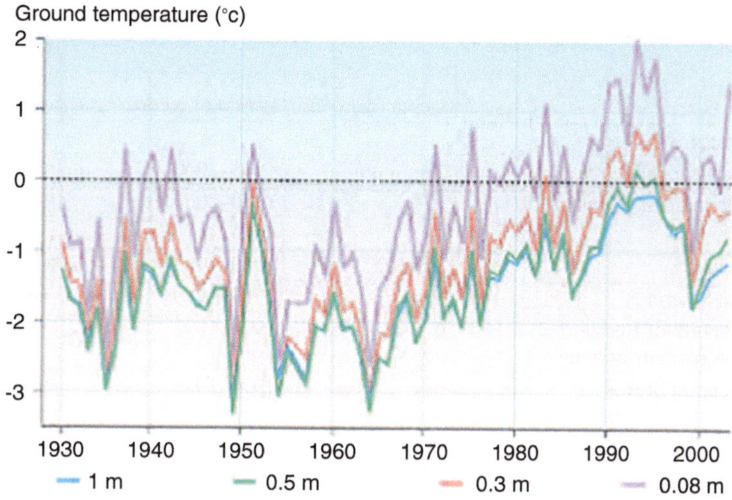

Fig. 12.1 Simulated mean annual ground temperature at Bonanza Creek, Fairbanks, Alaska, from 1930 to 2003 (Romanovsky 2004)

Global Trends Network-Permafrost (GTN-P), we find that the MAGT warmed throughout the world's permafrost region, with values ranging from 0.1 to 0.7 °C per decade for the past several decades in high-mountain regions of the world, 0.3–0.7 °C per decade in arctic Alaska, and 1.0 °C per decade in North Victoria Land, Antarctica (Table 12.1) (Fig. 12.1).

Permafrost temperatures for several stations in Antarctica, the Arctic, and high-mountain regions are given in Table 12.2. Permafrost that is −3 °C or warmer is considered to be "warm" permafrost and is very susceptible to thawing from climate warming (Fig. 12.2).

12.4 Possible Impacts on Soil Properties

In the Arctic continued warming can be expected to increase the ice-free area. An increase in shrubs (Loranty and Goetz 2012) and net primary production of vascular plants has already been observed (Chapin et al. 1995), which will likely increase soil organic C and N. Thickening of the active layer, especially in areas of sporadic and isolated permafrost, has resulted in an increase in thermokarst (Lenz et al. 2013) and likely will result in an increase in redoximorphism and pervection (see Table 8.5). In the High Arctic increased precipitation may reduce salinization. These same changes can be expected in areas with alpine cryosols.

Table 12.2 Recent trends in permafrost temperature

Region	Depth (m)	Period of record	Permafrost temperature change[a] (°C)
United States			
Trans-Alaska pipeline route	20	1983–2000	+0.6 to +1.5
Barrow Permafrost Observatory	15	1950–2001	+1
Russia			
East Siberia	1.6–3.2	1960–1992	+0.03/year
Northwest Siberia	10	1980–1990	+0.3 to +0.7
European north of Russia, continuous permafrost zone	6	1973–1992	+1.6 to +2.8
European north of Russia, discontinuous permafrost zone	6	1970–1995	Up to +1.2
Canada			
Alert, Nunavut	15–30	1995–2000	+0.15/year
Northern Mackenzie Basin, Northwest Territories	28	1990–2000	+0.1/year
Central Mackenzie Basin. Northwest Territories	15	1985–2000	+0.03/year
Northern Quebec	10	Late 1980s–mid-1990s	−0.1/year
Norway			
Juwasshoe, southern Norway	~5	Past 60–80 years	+0.5 to +1.0
Svalbard			
Janssonhaugen	~5	Past 60–80 years	+1 to +2

Romanovsky et al. (2002); in ACIA (2013)
[a]Temperature change over period of record, unless otherwise noted

In Antarctica warming will increase the ice-free areas along the western Antarctic Peninsula (WAP) and in coastal East Antarctica (Table 12.3). Increases in vascular plant cover have already been reported, especially along the WAP (Hill et al. 2011). Exotic species have already been introduced to Antarctica (Wodkiewicz et al. 2013) and can be expected to replace native species. The increase in plant cover will result in an increase in soil organic C and N (Bokhorst et al. 2007). Thawing of permafrost has resulted in active-layer detachment slides and thermokarst in the South Shetland Islands (Vieira et al. 2008; Bockheim

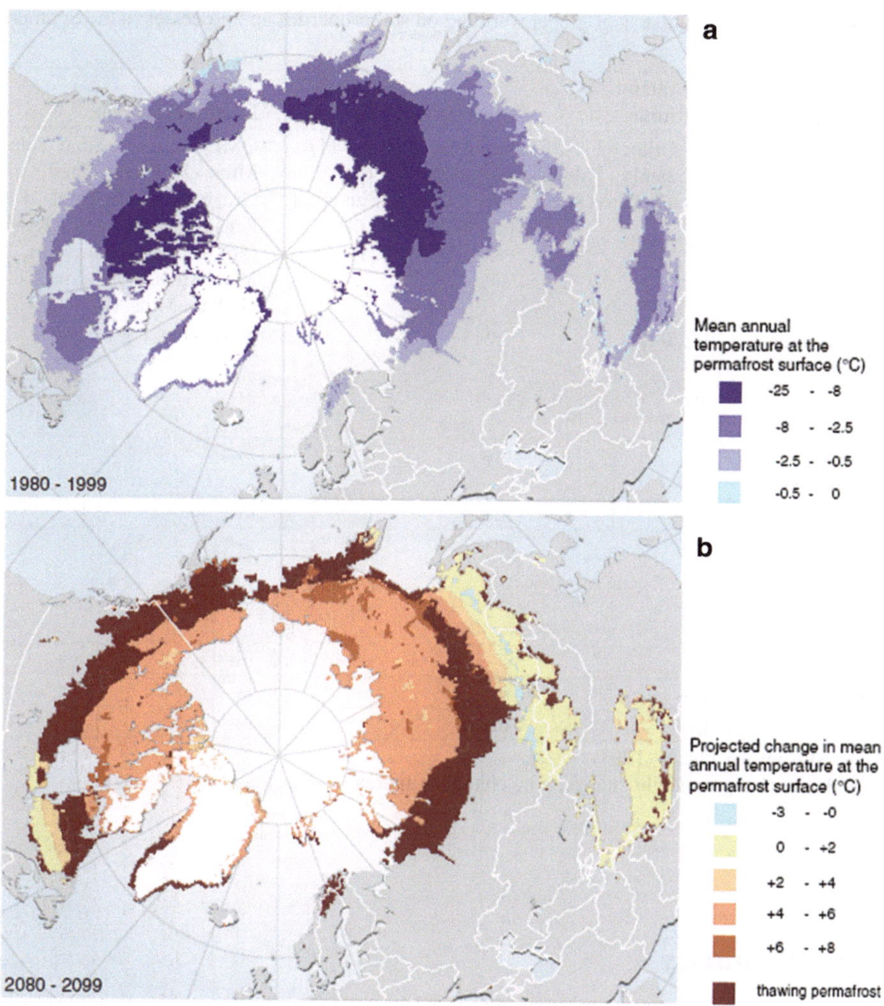

Fig. 12.2 Current and projected permafrost temperatures for the Northern Hemisphere (Romanovsky et al. 2007)

et al. 2013). We can expect increased redoximorphism and pervection in a warming climate along the WAP and in coastal East Antarctica. Warming effect on soils have been observed in the interior mountains of Antarctica, including reactivation of ice- and sand-wedge polygons, leaching of salts from soils, an increase in the area of the hyporheic zone, and a loss in semi-permanent snow patches leaving nivation hollows.

Table 12.3 Potential impacts of global warming on soil properties and processes in the Southern Circumpolar Region

Bioclimatic zone	W. Antarctica Maritime	E. Antarctica Maritime	Interior Mountains
Regions	So. Orkney I., So. Shetland I., Palmer Archipelago (8)	Schirmacher Oasis (1), Molodezhnaya (2), Ingrid Christensen coast (3), Windmill I. (4)	QML Mtns. (1), Scott-Tula Mtns. (2), Prince Chas., Grove Mtns. (3), Thiel-Pensacola Mtns. (5a), TAM (5b), Ellsworth Mtns. (6), MBL Mtns. (7), Palmer-Graham Mtns. (8)
Ice-free area (km^2)	1,200 (2.4 %)	1,245 (2.5 %)	47,055 (95.1 %)
Change in ice-free area	m+	m+	s+
Primary production	m+	m+	s+
Respiration	m+	m+	s+
Soil organic C (%)	m+	m+	s+
Depth to ice cement	m+	m+	s+
Salinization	o	s−	m−
Rubification	m+	m+	s+
Pervection	m+	m+	s+

Adapted from Bockheim (1993)
m moderate change, *s* slight change, *o* no change, + increase, – decrease

12.5 Summary

The high-latitude and high-elevation cryosols are experiencing some of the greatest warming over the past several decades of anywhere on Earth. This has resulted in increases in active-layer thickness in low Arctic and alpine cryosols and an increase in permafrost temperatures. There have already been reports of changes in soil properties and processes and continued changes can be expected (Fig. 12.3).

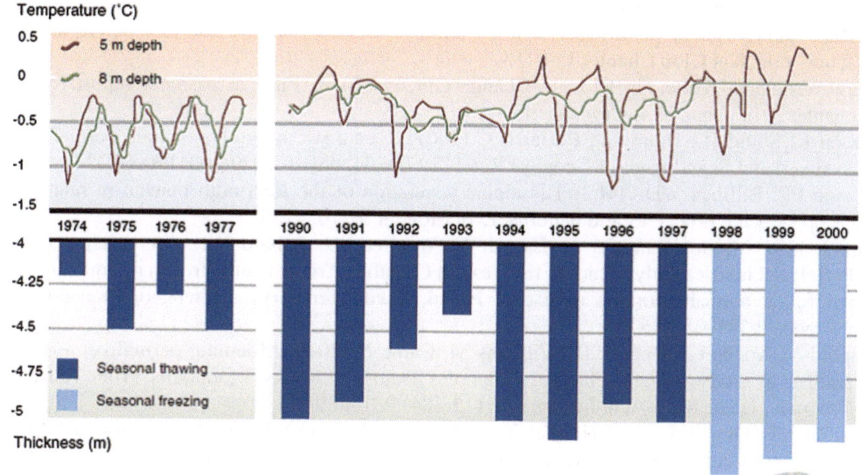

Fig. 12.3 Permafrost temperatures and active-layer thickness during 1974–1977 and 1990–2004 measured in a borehole at the "Cosmostation" permafrost observatory, 3,300 m above sea level, northern Tien Shan Mountains of central Asia (Marchenko et al. 2007)

References

ACIA (2013) Arctic climate impact assessment, Chapter 6. Cryosphere and hydrology, pp 183–242

Bockheim JG (1993) Global change and soil formation in the Antarctic region. In: Gilichinsky, DA (ed) Joint Russian-American seminar on cryopedology and global change. Russian Academy of Sciences, Pushchino (Moscow region), pp 132–140

Bockheim JG, Walker DA, Everett LR (1998) Soil carbon distribution in nonacidic and acidic tundra of arctic Alaska. In: Lal R, Kimble JM, Follett RF, Steward BA (eds) Soil processes and the carbon cycle. CRC Press, Boca Raton, pp 143–155

Bockheim JG, Vieira G et al (2013) Climate warming and permafrost dynamics on the Antarctic Peninsula. Glob Planet Change 100:215–223

Bokhorst S, Huiskes A, Convey P, Aerts R (2007) Climate change effects on organic matter decomposition rates in ecosystems from the martime Antarctic and Falkland Islands. Glob Change Biol 13:2642–2653

Chapin FS III, Shaver GR, Giblin AE, Nadelhoffer KJ, Laundre JA (1995) Responses of arctic tundra to experimental and observed changes in climate. Ecology 76:694–711

Christiansen HH (2004) Meteorological control on interannual spatial and temporal variations in snow cover and ground thawing in two northeast Greenlandic Circumpolar-Active-layer-Monitoring (CALM) sites. Permafr Periglac Process 15:155–169

Christiansen HH, Etzelmuller B, Isaksen K, Juliussen H (14 co-authors) (2010) The thermal state of permafrost in the Nordic area during the International Polar year 2007–2009. Permafr Periglac Process 21:156–181

Feuillet T (2011) Statistical analyses of active patterned ground occurrence in the Taillon Massif (Pyrénées, France/Spain). Permafr Periglac Process 22:228–238. doi:10.1002/ppp.726

Fukui K, Fujii Y, Mikhailov N, Ostanin O, Iwahana G (2007) The lower limit of mountain permafrost in the Russian Altai Mountains. Permafr Periglac Process 18:129–136. doi:10.1002/ppp.585

Guglielmin M, Cannone N (2011) A permafrost warming in a cooling Antarctica? Clim Change 111(2):177–195. doi:10.1007/s10584-011-0137-2

Hill PW, Farrar J, Roberts P, Farrell M, Grant H, Newsham KK, Hopkins DW, Bardgett RD, Jones DL (2011) Vascular plant success in a warming Antarctic may be due to efficient nitrogen acquisition. Nat Clim Change 1:50–53

Intergovernmental Panel on Climate Change (IPCC) (2013) Fifth assessment report: climate change 2013. Chapter 4: Cryosphere

Isaksen K, Sollid JL, Holmlund P, Harris C (2007) Recent warming of mountain permafrost in Svalbard and Scandinavia. J Geophys Res 112:F02S04. doi:10.1029/2006JF000522

Johnson PL, Billings WD (1962) The alpine vegetation of the Beartooth plateau in relation to cryopedogenic processes and patterns. Ecol Monogr 32:105–135

Lenz J, Fritz M, Schirrmeister L, Lantuit H, Wooller MJ, Pollard WH, Wetterich S (2013) Periglacial landscape dynamics in the western Canadian Arctic: results from a thermokarst lake record on a push moraine (Herschel Island, Yukon Territory). Palaeogeogr Palaeoclimat Palaeoecol 381–382:15–25

Leopold M, Voelkel J, Dethier D, Williams M, Caine N (2010) Mountain permafrost – a valid archive to study climate change? Examples from the Rocky Mountains Front Range of Colorado, USA. Nova Acta Leopoldina 112(384):281–289

Li R, Zhao L, Ding Y-J, Wu T-H, Xiao Y, Du E, Liu G-Y, Qiao Y-P (2012) Temporal and spatial variations of the active layer along the Qinghai-Tibet Highway in a permafrost region. Chin Sci Bull 57:4609–4616. doi:10.1007/s11434-012-5323-8

Loranty MM, Goetz SJ (2012) Shrub expansion and climate feedbacks in Arctic tundra. Environ Res Lett 7. doi:10.1088/1748-9326/7/1/0115005

Marchenko SS, Gorbunov AP, Romanovsky VE (2007) Permafrost warming in the Tien Shan Mountains, Central Asia. Glob Planet Change 56:311–327

Mazhitova G, Malkova G, Chestnykh O, Zamolodchikov D (2004) Active-layer spatial and temporal variability at European Russian Circumpolar-Active-Layer-Monitoring (CALM) sites. Permafr Periglac Process 15:123–139

Osterkamp TE (2007) Characteristics of the recent warming of permafrost in Alaska. J Geophys Res 112. doi:10.1029/2006JF000578

Romanovsky VE (2004) Geophysical Institute, University of Alaska, Fairbanks. http://www.eoearth.org/View/article/1551851. Accessed 26 Sept 2014

Romanovsky VE, Gruber S, Instanes A, Jin H, Marchenko SS, Smith SL, Trombotto D, Walter KM (2007) Frozen ground. In: Global outlook for ice and snow. United Nations Environment Programme, Nairobi

Sharkhuu N (2003) Recent changes in the permafrost of Mongolia. In: Phillips M, Springman S, Arenson L (eds) Permafrost, proceedings of the eighth international conference, 21–25 July, 2003, Zurich, Switzerland, vol 1. Balkema, Lisse, pp 1029–1034

Tarnocai C, Nixon FM, Kutny L (2004) Circumpolar-Active-Layer-Monitoring (CALM) sites in the Mackenzie Valley, northwestern Canada. Permafr Periglac Process 15:141–153

Tarnocai C, Canadell JG, Schuur EAG, Kuhry P, Mazhitova G, Zimov S (2009) Soil organic carbon pools in the northern circumpolar permafrost region. Global Biogeochem Cycle 23. doi:10.1029/2008GB003327

Turner J, Bindschadler RA, Convey P, Di Prisco G, Fahrbach E, Gutt J, Hodgson DA, Mayewski PA, Summerhayes CP (2009) Antarctic climate change and the environment. Scientific Commissions on Antarctic Resolution, Cambridge

Vieira G, López-Martínez J, Serrano E, Ramos M, Gruber S, Hauck C, Blanco JJ (2008) Geomorphological observations of permafrost and ground-ice degradation on deception and Livingston Islands, Maritime Antarctica. In: Kane D, Hinkel K (eds) Proceedings of the 9th international conference on permafrost, 29 June–3 July 2008, Fairbanks, Alaska, Extended Abstracts, vol 1. University of Alaska Press, Fairbanks, pp 1839–1844

Wodkiewicz M, Galera H, Chwedorzewska KJ, Gielwanowska I, Olech M (2013) Diaspores of the introduced species Poa annua L. in soil samples from King George Island (South Shetlands, Antarctica). Arctic Antarct Alp Res 45:415–419

Chapter 13
Management of Cryosols

13.1 Antarctica

13.1.1 Land Use

Antarctica is protected from development by the Antarctic Treaty (AT) (http://www.ats.aq/e/ats.htm). The AT has been signed by all 12 of the original signatory countries but also by 38 countries that accede to the AT. The AT limits the kind of activities can take place on the continent and requires environmental impact statements for small-scale scientific activities as well as major activities such as establishment of scientific bases. The primary use of Antarctica today is science, which for the US program includes astrophysics and space sciences, earth sciences, glaciology, ocean and atmospheric sciences, organisms and ecosystems, and integrated system science. Only 0.35 % (49,500 km^2) of Antarctica is ice-free. As most human activities are concentrated in relatively small ice-free areas, particularly in the Ross Sea region and Antarctic Peninsula, the potential for adverse human impacts on the soil landscape is great (O'Neill et al. 2014).

Tourism is becoming an increasingly important activity in Antarctica, including cruise ships from South America and mountaineering expeditions working out of the Heritage Range of the Ellsworth Mountains. Tourism in Antarctica has increased by 344 % in the past 13 years (International Association of Antarctic Tour Operators 2013).

J.G. Bockheim, *Cryopedology*, Progress in Soil Science,
DOI 10.1007/978-3-319-08485-5_13

13.1.2 Land Degradation

The main land degradation concerns in Antarctica include landscape modification as a result of construction activities, geotechnical studies, and road construction; disturbance to soil communities; local pollution from hydrocarbon spills; waste disposal; and the introduction of alien species (O'Neill et al. 2014).

13.2 Arctic

13.2.1 Land Use

The Arctic has a more agreeable climate than Antarctica and has sustained peoples for at least 13,000 year. Before the twentieth century, the primary activity in the Arctic was subsistence living by about 40 different ethnic groups that included fishing, reindeer herding, whaling, sealing and related activities. From an economic standpoint, the first major activity in the Arctic was gold mining. Gold mining began in Siberia and the Russian Far East in the 1830s. The gold output in Russia peaked in 1855 and declined steadily until World War I. After the war gold and silver mining were renewed during the tragic gulags. The Klondike Gold Rush in the Yukon Territory of Canada occurred between 1896 and 1899 and brought an estimated 100,000 prospectors to this permafrost region.

Petroleum and gas extraction became the major activities in the Arctic beginning in the 1920s in Canada's Northwest Territories and in the 1960s on Alaska's North Slope, the Yalmalo–Nenets region of Russia, and the Mackenzie Delta of Canada. More recent discoveries have been made in Greenland (2001) and northern Norway (2012). Although most of these operations are offshore on the continental shelves, pipelines have been constructed across land to carry the petroleum products to deep-water ports.

There is a long history of agriculture in the Fennoscandian and Russian Arctic. Agriculture is a relatively small industry in the Arctic and is composed mainly of cool-season forages, cool-season vegetables, and small grains and raising traditional livestock (cattle, sheep, goats, pigs, poultry, and horses) (ACIA 2013). Key agricultural areas include the southern NWT; the Matanuska Valley, Alaska; Rovaniemi, Finland; Novosibirsk, Yakutsk, Vorkuta, and Magadan, Russia; the three northern counties of Norway; coastal Iceland and Greenland. Figure 13.1 shows agriculture on permafrost near Bethel, western Alaska.

Although the Arctic has attracted tourists since the early 1800s, it has become an important industry in the past 15 years. Many of the larger tour operators adhere to a Code of Conduct in order to protect the environment and cultures of the Arctic.

Fig. 13.1 Farming on permafrost near Bethel, western Alaska (http://www.neatorama.com)

13.2.2 Land Degradation

Land degradation issues in the Arctic include chemical pollution from mining, oil and gas extraction, and long-distance transport of persistent organic pollutants (POPs), heavy metals, radionuclides, and acidifying gases; removal of taiga forests; loss of biodiversity; removal of sand and gravel for access roads and pads for pumping stations; and flooding from hydroelectric development (UNEP 1997). Of major concern is the impact of these activities on indigenous peoples residing in the Arctic. Figure 13.2 shows the Meadowbank gold mine near Baker Lake, Nunavut, Canada, an area underlain by permafrost. Figure 13.3 shows an oil spill near Usinsk in the Komi Republic of Russia. Figure 13.4 shows collapse of a road in permafrost in interior Alaska.

13.3 Alpine Regions with Permafrost

13.3.1 Land Use

The high mountain regions have been used primarily for mining, recreation, limited agriculture, and highway-railway construction. Peter the Great issued a "mining privilege" in 1719 so that gold and silver mining took place as early as 1733–1735 in the Altai Mountains. Mining also accounts for the location of several of the world's highest "villages," including La Rinconada, Peru (5,100 m; gold), Pascu–Lama,

Fig. 13.2 The Meadowbank gold mine near Baker Lake, Nunavut, Canada (Photo courtesy of Reuters/Euan Rocha)

Fig. 13.3 Oil spill near Usinsk, Komi Republic, Russia (Photo courtesy of Dmitry Lovetsky/Associated Press)

Argentina (4,500 m; gold), El Aquilar, Argentina (4,895 m; lead, zinc, and platinum), and Potosi, Bolivia (4,091 m; silver). Some of these "villages" occur in areas of sporadic or isolated permafrost. At an elevation of 5,100 m, Wanquan, China serves the Qinghai-Tibet railway and has a military installation. Murghab in the Pamir Mountains of Tajikistan is at an elevation of 3,650 m and has 4,000 residents that deal primarily in trade. There are a few villages in mountains above the Arctic Circle and above 500 m elevation that have permafrost, such as Anaktuvuk Pass in

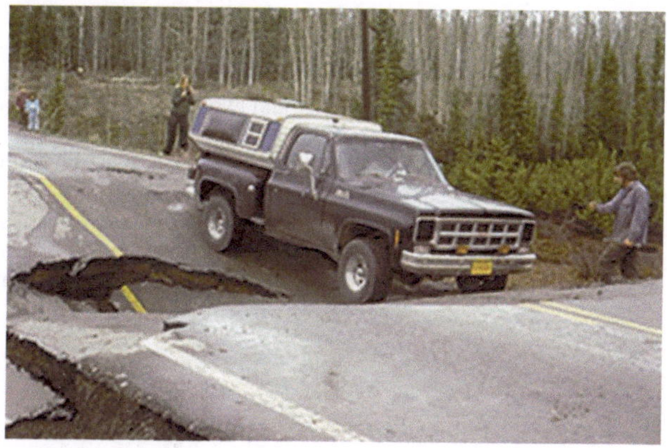

Fig. 13.4 Melting of permafrost causes a road to collapse in Alaska (Photo courtesy of Federal Highway Administration, 2010)

Alaska's Brooks Range. Tourism is a major industry in the European Alps and will become important in the Caucasus Mountains following the 2014 Winter Olympics.

13.3.2 Land Degradation

Global warming is the key environmental issue in high-mountains regions and was considered in the previous chapter. Other environmental concerns are overgrazing, deforestation in the subalpine, and landslides and flash floods from melting of permafrost. Mining and mining waste disposal are issues in the Fennoscandian and Andes Mountains. After-effects from radioactive fallout from the Chernobyl disaster continue to threaten alpine ecosystems in Eastern Europe.

13.4 Summary

The key land uses in Antarctica are establishment of scientific bases and tourism. These activities have had localized impacts in terms of petroleum spills and release of toxic chemicals.

In the Arctic key land uses are subsistence living, oil and gas extraction, mining, and more recently tourism. Land degradation in the Arctic is manifested by spills of petroleum products from pipelines, soil and water contamination from mining, removal of timberline forests, and removal of sand and gravel for access roads and pads for pumping stations. Key land uses in areas with mountain permafrost include mining, recreation, limited agriculture, and highway-railway construction. Land degradation in alpine regions with permafrost include erosion and mass wasting from cutting of subalpine forests, flash floods from melting of permafrost, soil and water contamination from mining and mining waste disposal, and the after-effects from radioactive fallout from the Chernobyl disaster.

References

ACIA (2013) Chapter 6: Cryosphere and hydrology. In: Arctic climate impact assessment. AMAP, Oslo, pp 183–242

International Association of Antarctic Tour Operators (IAATO) (2013) Tourism statistics. http://iaato.org/tourism-statistics. Accessed 17 Mar 2014

O'Neill T, Aislabie J, Balks MR (2014) Human impacts on soils. In: Bockheim JG (ed) Soils of Antarctica. Springer, New York

United Nations Environment Programme (1997) Global environment outlook-1

Chapter 14
Cryosol Databases

14.1 Database Development

Soil databases are collections of soil information organized in systematic form in an electronic environment. They include both spatial databases, which contain soil information that can be displayed as soil maps, and point databases, which contain morphological, physical and chemical data for a pedon at a specific location. In addition, numerous soil databases contain monitored data on soil temperatures, soil moisture, and active and thaw layer depths. Since the soil databases are in an electronic form, they are very useful for various interpretations, scaling up information such as carbon concentrations and carbon stocks, and providing basic information for modeling.

At the present time there are spatial databases for cryosols in Canada (Agriculture and Agri-Food Canada 2014a), the USA (Soil Survey Division 2014a), Russia (Fridland 1988), Greenland (Jakobsen and Eiby 1997). These spatial databases present information at scales of 1:1,000,000–1:2,000,000 (small-scale databases). One of the most comprehensive spatial databases is the Northern and Mid Latitude Soil Database (European Union, Joint Research Centre 2014). This database, which is also a small-scale database, contains soil information for North America, Greenland, Europe and the northern part of Asia. Unfortunately, there are very few large-scale spatial soil databases available for permafrost areas.

Currently, no pedon database (a point database) has been developed specifically for cryosols. Databases such as the U.S. and Canadian pedon databases (Soil Survey Division 2014b; Agriculture and Agri-Food Canada 2014b) contain pedons of cryosols, but the number of such pedons is very small. A summary of cryosol databases is given in Table 14.1.

J.G. Bockheim, *Cryopedology*, Progress in Soil Science,
DOI 10.1007/978-3-319-08485-5_14

Table 14.1 Cryosol databases

Type	Area included	Data included	Agency	Website	Contact
Spatial	Canada		Centre for Land & Biological Resour. Res.		C. Tarnocai
Spatial	USA	Web soil survey (SSURGO, STATSGO)	Natural Resources Conserv. Serv.		J. Hempel
Spatial	Circumarctic	Northern and mid-latitude database	Cryosol Working Group		C. Tarnocai
Spatial	maritime Antarctica	Soil maps of selected areas			C.E.G.R. Schaefer
Spatial	McMurdo Dry Valleys	Soil maps TAM, MDV, Wright V.	Landcare Research, New Zealand		M McLeod
Pedon	USA	Official soil series descriptions, soil characterization data	Natural Resources Conserv. Serv.		J. Hempel
Pedon	Canada	Official soil series descriptions, soil characterization data	Agriculture & Agri-Food Canada		C. Tarnocai
Pedon	Antarctica	soil descriptions, analytical data	NSIDC		J.G. Bockheim
Pedon	Antarctica	Soil descriptions, analytical data	NSIDC		I.B. Campbell
Pedon	Antarctica	Soil descriptions, analytical data			C.E.G.R. Schaefer
Pedon	Europe	Soil descriptions, analytical data			

References

Agriculture and Agri-Food Canada (2014a) Web maps. Canadian Soil Information Service. http://sis.agr.gc.ca/cansis/publication/webmaps.html. Accessed 17 Mar 2014

Agriculture and Agri-Food Canada (2014b) Soil name and layer data for Canada. Canadian Soil Information Service. http://sis.agr.gc.ca/cansis/soils/provinces.html. Accessed 17 Mar 2014

European Union, Joint Research Centre (2014) Soil data and information systems. Northern and Mid Latitude Soil Database. http://eusoils.jrc.ec.europa.eu/library/esdac/Esdac_DetailData2.cfm?id=49. Accessed 17 Mar 2014

Fridland VM (1988) Pochvennaya karta RSFSR (Soil map of the Russian republics). Dokuchayev Soils Institute, Moscow (scale 1:2,500,000)

Jakobsen BH, Eiby A (1997) A soil map of Greenland. In: 2nd International conference on cryopedology, Syktyvkar, Russia
Soil Survey Division (2014a) Web soil survey. http://websoilsurvey.sc.eov.usda.gov/. Accessed 17 Mar 2014
Soil Survey Division (2014b) Official soil series descriptions. http://soilseries.sc.egov.usda.gov/. Accessed 17 Mar 2014

Index

J.G. Bockheim, *Cryopedology*, Progress in Soil Science,
DOI 10.1007/978-3-319-08485-5

E

F

G

H

I

J

K

L

M

N

O

P